Water in North American Environmental History

Water in North American Environmental History offers 25 case studies that explore the range of uses and perceptions of water throughout Canadian, Mexican, and United States history.

Water has served a myriad of purposes historically as human sustenance, agricultural irrigation, sanitation, fire protection, military defense, power generation, transportation, and much more. Water and its uses provide an excellent entrée into the study of humans and the environment, not only because water is a vital resource for life, but also because water as a medium is so intimately woven into the everyday experiences of humans and into society's economic, political, and social fabric. A North American perspective is not representative of the world's water use, but it is an area with a linked history and many overlapping human and environmental features and concerns. With a continental perspective, the book explores many disparate topics without being confined to the history and experiences of just one country. The chapters are short but descriptive, and departure points for what they tell us about the human experience in dealing with water and the environmental implications of water use. The text leads students to consider water in relation to society, and to the past.

The book will be of interest to students of environmental history, geography, and the environmental sciences.

Martin V. Melosi is Cullen Professor Emeritus of History and Founding Director of the Center for Public History at the University of Houston, USA. He studies environmental and urban history and energy history.

Themes in Environmental History

Themes in Environmental History is a series of books aimed at 2nd- and 3rd-year undergraduate students and postgraduate students in the fields of history and environmental studies. The collection covers key areas of environmental history from across the globe, running from 500 CE to the present day. These books bring together chapters on the historiography of the field and the new research that is being done to move the field forward, making engaging reading for students. Topics covered are varied and expansive and emphasize the importance of looking back at environmental history to date to understand where we are today.

Water in North American Environmental History
Martin V. Melosi

Disease and the Environment in the Medieval and Early Modern Worlds
Edited by Lori Jones

For more information about this series, please visit: https://www.routledge.com/Themes-in-Environmental-History/book-series/TIEH

Water in North American Environmental History

Martin V. Melosi

Routledge
Taylor & Francis Group

NEW YORK AND LONDON

Cover image: River operations at Murderers Bar during the California Gold Rush 1850s. Hand-colored woodcut. North Wind Picture Archives/Alamy Stock Photo.

First published 2022
by Routledge
605 Third Avenue, New York, NY 10158

and by Routledge
4 Park Square, Milton Park, Abingdon, Oxon, OX14 4RN

Routledge is an imprint of the Taylor & Francis Group, an informa business

© 2022 Taylor & Francis

The right of Martin V. Melosi to be identified as the author of this work has been asserted in accordance with sections 77 and 78 of the Copyright, Designs and Patents Act 1988.

Library of Congress Cataloging-in-Publication Data
Names: Melosi, Martin V., 1947- author. Title: Water in North American environmental history / Martin V. Melosi.
Description: New York, NY : Routledge, 2022. |
Series: Themes in environmental history | Includes bibliographical references and index. |
Identifiers: LCCN 2021058355 (print) | LCCN 2021058356 (ebook) |
Subjects: LCSH: Water resources development--North America--History. |
Hydraulic engineering--North America--History. | Environmental engineering--North America--History. | Water-supply--North America--History. | Human ecology--North America--History. |
North America--Environmental conditions.
Classification: LCC TC422 .M45 2022 (print) | LCC TC422 (ebook) |
DDC 333.910097--dc23/eng/20220125
LC record available at https://lccn.loc.gov/2021058355
LC ebook record available at https://lccn.loc.gov/2021058356

ISBN: 978-0-367-48554-2 (hbk)
ISBN: 978-0-367-48553-5 (pbk)
ISBN: 978-1-003-04162-7 (ebk)

DOI: 10.4324/9781003041627

Typeset in Bembo
by MPS Limited, Dehradun

To Scout, Charles, Tom K., James, Jan, Jonathan, Graham, Terri, Kim, Tom M., Jordan, Julie, Joe, Jeff, and all of my other students. I have learned so much from you.

Contents

Acknowledgments

This is the first book I have written (and maybe the last?) since retiring from the University of Houston in 2018. This experience has been unusual because I have conducted the research and have written the whole thing largely in my home office because of the Covid-19 pandemic. Although I have remained healthy, the strategy for completing this study depended very much on these circumstances. My research comes from my personal library, books purchased through Amazon, the Internet, and through Interlibrary Loan and the on-site book and journal collections at the MD Anderson Library at UH. Having decided to focus the study on North America, rather than simply drawing upon examples from the United States, was challenging but worth it. I think a perspective on water which crosses the borders of Mexico, the United States, and Canada added richness and perspective to the book-writing experience.

I appreciate input from reviewers selected by Taylor and Francis and the help from Kimberley Smith and Emily Irvine at the press. Who would have guessed that from my first correspondence with Kimberley that this book, under these circumstances, would emerge at all?

I also want to thank my wife Carolyn, who tolerated my constant chatter about the various water episodes, but also knew I was happy when banging out a new book on my computer. Only my granddaughters, Gianna and Angelina, could happily distract me from the work. I hope the results of this effort will be useful to a new generation of students, who find the book on their syllabi.

Martin V. Melosi

Introduction

In his book, *Water 4.0: The Past, Present, and Future of the World's Most Vital Resource*, engineer David Sedlak talks about "the hidden world of water" where "the miles of pipes that bring water into our homes from distant locations, the treatment plants that ensure the wastes we pour into the sink and flush down of toilet don't pollute the local river, and the network of storm drains that keep the rain from flooding our homes continue to operate silently, day and night, whether or not we are aware of their existence" (Sedlak, 2014, vii).

This pretty well describes the fundamental water systems that operate in modern cities. But water is not just hidden or invisible in our world. The oceans, lakes, rivers, ponds, and streams are visible in many places as is rain, sleet, and snow. Glaciers—while they still exist in considerable numbers—are not so close to most people, but we know they are there. As environmental historian John McNeill observed, "Earth is the water planet, the only place in our solar system where water exists as liquid" (McNeill, 2000, 119). While about 97 percent of our water is saltwater, and the rest fresh, our hydrosphere makes up about 71 percent of the earth's surface.

We know that people need fresh water to survive, just as much as we need oxygen to breathe. On a global scale, consumption of water since the 1980s has increased one percent per year; global population growth also runs at about the same rate. But in the twentieth century, water withdrawals increased enormously over previous centuries (probably more in Europe than in North America). Asia remained the hands-down biggest consumer in the late twentieth century with withdrawals greater than all other continents combined, largely because of the size of its population and the presence of about one-third of the world's streamflow.

We sometimes forget how versatile water can be and are not always mindful of the multitude of ways water influences our lives, be it hidden or visible. Nor are we aware of the extensive environmental implications of using water in the home, in the workplace, or in the community at large. I have stated elsewhere that "Water as a medium...serves myriad purposes—human sustenance, agricultural irrigation, sanitation, fire protection, military defense, power generation, transportation, and more—and thus contested uses must be

DOI: 10.4324/9781003041627-1

understood and explained to obtain a proper context in which to evaluate wants and needs. Underlying water's uses are cultural traditions, various hierarchies, and social perceptions and conditions—all influencing control and allocation" (Melosi, 2011, xiii).

This book deals with water in environmental history, especially fresh water and, in some instances, coastal/marine water issues. (Blue humanities is a newly emerging field dealing with the relationship between humans and the seas and thus must be included to some degree in the book.) It is composed of twenty-five, what I call, water episodes in 8 largely chronological sections covering precontact years to the present in North America. The book is meant for use in the classroom as a way of introducing students to a variety of uses and perceptions of water. These water episodes are short and not comprehensive learning vehicles in themselves. They are meant as a departure point for discussion and the basis for starting conversations about the topics of the episodes and what they might tell us about larger water issues in our history, including the human and environmental implications of water use.

Rather than an American perspective only, the book covers North America—Mexico, the United States, and Canada. Alice Outwater, an environmental engineer, argued that "Water is the blood of the land—always in motion, from the rain to the mountaintops, through the forests and plains to the sea, and so to the clouds again. And yet, on the North American continent, the natural cycle has been changed in a number of ways. As a result, water is no longer able to clean itself naturally, and despite our best legislative efforts our waterways are still impaired" (Outwater, 1996, xii). Writing in 1996, Outwater was frustrated about the current state of the continent's water, and not that things have improved (some have gotten worse) but her reflections tell only part of a complex story about North America's relationship to water.

A continental perspective allows us to explore many topics while going beyond thinking about water episodes within a single country or a single set of experiences. In some cases, since water flows, political boundaries are intellectually restrictive. In others, the episodes can concentrate on a particular country, but raise questions that relate to all three. By no means is a North American perspective representative of the world's water use, but it is an area with a common (or sometimes aggregate) history and many overlapping physical features, while also including human diversity, differing water policies, and unique practices useful for comparative purposes.

My intention is to cover as many themes and topics as possible in this book, but page limits dictate constraints. There are many more water episodes to choose from than are included in the book. It was important to include topics that are varied, but also fit within a broad chronology and are disseminated among the three countries. This is no easy task, and some people will question why this topic or that was or was not included. My primary goal, however, is to avoid making the water episodes one-dimensional or single-themed, rather address several different themes in each case where possible.

Guiding my hand is an emphasis on an environmental history of water that gives attention to technical, political, economic, and social issues. I see environmental history writ large as the interactions of humans with their environment, yet I try to avoid so-called declensionist narratives, that is, where humans are simply and exclusively agents of harm in the settings where they live, work, and play.

I am particularly influenced by what has been labeled as an envirotech approach to history, where environmental history overlaps with the history of technology. Some scholars conclude that "nature and technology, and the ways in which we understand the two, have become increasingly entangled blurring boundaries that once seemed so clear" (Reuss and Cutcliffe, 2010, 6). This is not a new departure in history because anthropologists long have studied how various cultures relate to nonhuman nature. Aside from its potable use, water operates as a medium for transportation, a source of energy (solar, mechanical, etc.), a border or barrier between lands, a tool to do work, and an instrument of conveyance of disease, medicine, or pollution. It can create an environmental footprint through flooding, flowing along inconsistent or uncomfortable paths (for humans, that is), or through the destruction of hurricanes and other extreme forms of weather.

In these instances and more, we can view water through the concept of envirotechnical systems, where, as historians Sara B. Pritchard and Carl A. Zimring argued, environmental and technological systems "both shape and are shaped by each other" (Pritchard and Zimring, 2020, 9–10). They added, "By thinking about technology and the environment at the same time, we challenge clean divisions between environmental and technological processes, and instead assume that technical systems are never entirely closed systems" (Pritchard and Zimring, 2020, 15–16). To avoid viewing envirotechnical systems as inflexible is to make sure they are not treated as self-contained or autonomous. To do so would put people and what people do outside the process. Questions of race, gender, and class, as well, cannot be ignored.

Before one can begin to appreciate how water is utilized and what impact its use might have on the environment, we need to understand its basic physical properties. We all know that water can be liquid, solid, or gaseous, giving it infinitely more complexity than many other resources. Robert Glennon, professor of water law at the University of Arizona, puts this quite dramatically when he stated, "Water is the essence of life, the core of chemistry, the prime component of the human body…Without it, life ceases. With it, life can flourish" (Glennon, 2002, 13).

Fresh water is a resource with demand on the rise but limited in access to many. A common and oft-repeated statement in recent years is that "Water is the Next Oil." Nothing could be farther from the truth. "Water is the

Next Water" because of its longstanding historical role as the most essential natural resource (along with air) predating by centuries the fossil fuel era. The comparisons between water and oil, of course, are understandable. Oil has become scarcer and more expensive. The idea of "peak oil" had been dogma in the past, that is, accessible supplies of oil have been exploited and thus the world faces a downward spiral—fast or slowly depending upon who you believe—to the bottom of that energy barrel. Fracking techniques and new assessments of oil and gas supplies have changed—or at least adjusted the peak oil theory. The idea of a "fresh water crisis" or "peak water" also has captured the imagination of some, but that concern (or hysteria) has its own complications and overreactions (See Gleick, 2009, 1–16).

For years, scientists and others have debated how to understand water as a resource. Is it renewable or nonrenewable? Traditionally water has been regarded as renewable, at least in the sense that hydrologic cycles circulate it with very little lost in the process. Running water (flowing rivers) is often regarded as a "fugitive resource" like fish and wildlife, especially in efforts to define property rights for obvious reasons. Only groundwater is likely to be viewed as nonrenewable, especially if aquifers cannot be sufficiently recharged. Groundwater also has been regarded as a "common pool resource" that may be owned by a surface property owner under which the groundwater collects (Rogers, 1993, 76–77).

Fresh water can be trapped in glaciers, ice caps, permanent snow cover, or deep aquifers. Only about one percent of the earth's water is easily available for ecosystems and humans, but geography (and political control) may dictate who can get it and who cannot. By 2004, 31 countries (mostly in Africa and the Near East) confronted water stress or scarcities. Seventeen additional countries (with over 2 million people) were water-poor. Of that one percent is available fresh water, however not all of it is reliable drinking water.

Sorting out our best understanding about water as a physical resource in line with sustainable practices has led to claims of a "fresh water crisis" in the world. Jared Diamond argued, "…fresh water can be made by desalinization of seawater, but that costs money and energy, as does pumping the resulting desalinized water inland for use. Hence, desalinization, while it is useful locally, is too expensive to solve most of the world's water shortages"(Diamond, 2005, 490).

But "crisis" is a loaded term that has to be used clearly and accurately. While shortages of fresh water have become more common, in many cases the question also has to do with unsustainable practices that deplete water supplies faster than they are replenished, pollute water beyond reasonable usage, or limit control of water to a single party (or government).

There is a problem trying to include prodigious numbers of themes in one book—like the proverbial kid in a candy store—because there is so much to choose from with a ubiquitous resource like water. At the same time, the wide array of themes that the subject conjures up underpins the assertion that water shows up everywhere in our lives. Below is an array of important themes represented in the book (some find their way into multiple essays).

Water and Borders

Flowing rivers, especially, are physical features that create natural borders between peoples and nations. Such borders can be elusive since they can change course, dry up, or overflow beyond their banks. They can provide resources of their own such as water for drinking, a medium for fishing, or become a privy for waste. The Rio Grande River between the United States and Mexico is one of those borders. Historically, changes in ownership of lands due to conquest or other means often meant substantial changes in the control of rivers and streams. For example, the takeover of former Mexican lands by U.S. forces in the upper Rio Grande watershed during the Mexican–U.S. War in the 1840s had a variety of impacts: transfers of land grant commons to private owners or the U.S. Forest Service, replacement of communal irrigation districts with market-oriented conservancy districts, replacement of a subsistence economy with a market economy or industrial development, and other actions all influencing water use.

Borders between coastal waters and land pose their own issues, such as Toronto's effort to remake its harbor to improve its commerce and trade potential, and the growing competition by a variety of nations over coastal waters around the Grand Banks in order to control the immensely important cod-fishing grounds and the international markets for fish. Between the United States and Canada, along their western frontier, the competition was intense over controlling salmon fisheries and developing hydropower.

The Mexico–U.S. border also has seen the rise of maquiladoras—foreign-owned industries—that offered the hope of great economic prosperity for the two countries. However, they were built on the backs of low-wage earners and created vast amounts of pollution affecting both sides of the border. Likewise, the hydraulic fracking industry grew up on that same border, also bringing hope of economic boon, but generating great concern over the extensive use of local fresh water, creation of water-pollution problems, and production of earthquakes.

Water and Adaptation

An important theme, especially in the essays dealing with indigenous people in the precontact years, is the ability to adapt to their surroundings and their access to water. The Hohokam in the arid American Southwest built extensive irrigation canals and thus initiated a flourishing agricultural system. The Aztecs in Mexico constructed their capital city, Tenochtitlan, over a lake and swampy land, thus providing a water barrier between itself and its enemies. They utilized flowing water for transportation and trade and also built a substantial farming system. The Inuit in the Arctic confronted their hostile, icy environment in unique ways by mastering the knowledge of sea ice, weather cycles, and wildlife to build a sustainable society.

Water and Community

How societies choose to use and share water manifested itself in how certain societies were structured. The Hohokam viewed their irrigation systems as community-wide projects, with community-wide benefits. Spanish settlers and indigenous people after the Hohokam built acequias—community irrigation systems found in the villages and pueblos of New Mexico and the Southwest and also in northern Mexico and the Andes. Community use and control of irrigated water found its way into a body of laws, but laws were soon challenged by different notions of land and water ownership. The Inuit survived in the Arctic by establishing strong communal ties in confronting a very difficult—which white settlers believed was unlivable—environment. In Philadelphia in 1801, city leaders constructed the first community-wide water supply system, ushering in a period in the United States when public responsibility for services like water emerged as preferable to private development and control.

Water as Transportation

Using water as transportation was a quite obvious practice. Aztecs relied almost exclusively on water for making long trips and for commerce, lacking the wheel or animal power. The Inuit mastered the ability to move over ice, find game, trade, and avoid areas where the ice could not provide a good land bridge. They also were adept at traveling by canoe along the many waterways intersperse among the ice flows. The Erie Canal was a colossal public works project in its time in the early nineteenth century, conquering time for traveling great distances, opening western lands for settlement and trade, and communicating new ideas. A century later the Houston Ship Channel in southeast Texas linked an inland bayou with the Gulf of Mexico, creating one of the most important navigation and trading ports in the world. Toronto totally remade its harbor, more effectively connecting its coastline to seagoing trade and promoting greater industrial activity in the Canadian city. (New York City's experience provides a good comparison. See Kara Schlichting's *New York Recentered: Building the Metropolis from the Shore* (2019)).

Water and Energy

For many generations water was a necessary source of energy and also a necessary adjunct to some energy systems. Waterwheels depended on fast-moving water to function, thus were site-specific forms of energy generation that helped build industries such as cotton milling. They were important for changing moving water into mechanical power. Steam engines depended on water to create the steam necessary to run everything from gigantic turbines to steamboats and trains. Steam engines liberated producers of power from stationary sites to operating in almost any location and in a variety of scales.

Urban and industrial development in the nineteenth century depended on steam engines. Such technologies became major markets for wood and coal.

New methods of exploration and extraction offered ways to search and recover oil and gas in offshore locations, while hydraulic fracturing used water to force oil and gas out of deep fissures in the ground. We will see in the opening up of fisheries in Newfoundland in eastern Canada and salmon fisheries and canneries along the Fraser River in western Canada, the beginnings of large-scale commercial fishing that would take thousands upon thousands of pounds of fish each year. As environmental historian Richard White argued, "The flow of the river is energy, so is the electricity that comes from dams that block that flow. Human labor is energy; so are the calories stored as fat by salmon for their journey upstream" (White, 1995, ix).

Water as Commodity and Water as a Right

As seen in many of the essays, water can produce resources—such as fish—but also can be a commodity unto itself. The controversy over water privatization in recent years revolves, in part at least, around a concern among critics that water is becoming commodified, thus undermining the public's access and right to it. Probably it is more accurate to say that water has long been a commodity to be bought and sold, and what we are observing more recently as water privatization of urban water systems is a "recommodification."

A long-standing schism rests upon the ideas that water is as a public good versus water as a private commodity (or exclusive economic good). Such a paradigm is much too simple, a little too murky, and much too dichotomous. Water is a commodity whether it is managed and/or allocated by government or by a private business—or anything in between. But how that commodity is treated and for who benefits from it is an equally important issue.

A difficult question is whether we must accept fresh water as part of the commons, or subject it to ownership. The water episode on the Ogallala Aquifer in particular makes that point, as does the essay on Niagara Falls and others. Historically water has been treated as a public resource—such as the Southwestern irrigation systems—to be used or managed by communal entities; in other situations, the view of water as private property is often in line with concepts of property rights. As one study noted, "The struggle to control water is a struggle without end" (Melosi, 2011, xiii).

On another level, recent debates over water corporatism suggest that privatizing water systems is nothing more than giving in to the idea that water is a commodity to be bought and sold. In this way, the right to fresh water—and thus the right to health and well-being—falls out of the equation not only for people in the developed world but especially in the developing world where access to water is at best precarious for millions of people. The essay on Mexico's water management practices raises this issue most clearly.

The question of the right to clean water is, among other things, an environmental question. In the case of human water uses, it is a survival issue as well as a social justice concern for those without the means to acquire clean water at whatever price is set for it, if indeed access to water is even possible. According to water specialist Peter H. Gleick, "Until recently the question of whether individuals or groups have a legal right to a minimum set of resources, specifically water, and whether there is an obligation for states or other parties to provide those resources when they are lacking, has not been adequately addressed" (Gleick, 2000, 3).

The Flint water crisis exposed the vast chasm between those who were making decisions over providing water service for the city versus those forced to accept the consequences of those decisions. As Gleick stated, "Like most natural resources, water flows to those who have power" (Gleick, 2006, 117). Similarly, civic and state leaders who favored a policy of "levees-only" as a flood-control policy for the Mississippi River did not take into account the possible impacts of that policy on the poor or people of color as opposed to property owners along the river. Workers in the maquiladoras in Mexico had no role in setting policies to protect them from egregious exposures to hazardous wastes or to the presence of chemical pollutants in the rivers, streams, and groundwater in the area.

If there is a positive side to the question of water as a commodity—and this greatly depends on your vantage point—it has to do with the need to ask: Have we placed so little value on water—monetary or otherwise--throughout history that we treat it as if the supplies were endless? For those with ready access to water, the issue has to be "What is a fair price?" For those with limited or no access to clean water, the question transcends cost. It also is good to remember what professor of environmental engineering and city planning Peter Rogers stated about the United States: For over two hundred years "public goals concerning water policy have continually shifted because the nature of the problems has changed as the country has grown, both in size and in affluence" (Rogers, 1993, 1).

Water, Race, Gender, and Class

The cases of Flint, the maquiladoras, and others highlight some instances of environmental injustice associated with water use and related exposure to a variety of pollution problems. Some episodes, such as Flint, the maquiladoras, the fluoride controversy, and pollution in the Great Lakes through detergent phosphates have strong gender and age components. For example, the Flint crisis exposed the serious, and long-lasting, risk of lead poisoning from the Flint River in children. The fluoride controversy in large measure revolved around exposing children to a chemical treatment meant to reduce cavities, but one that led to a debate over the alleged "doping" of water. The biggest number of workers in the maquiladoras were women—who not only held low-paying jobs but were exposed to a range of toxic materials in the workplace and at home. A most overt form of racial injustice could be found in swimming pools in the United States and Canada where racial segregation and hostility to minorities was endemic.

Water as a Destructive Force

Almost every story in this book not only explores how humans use but also exploit water resources. There are many cases where this exploitation leads to scarcities or pollution, where use creates environmental footprints—some of which largely change the physical landscape itself. But water has a malevolent side as well. Hurricanes, such as Hazel, cause untold damage in its path. The floods along the Mississippi engulf, inundate, and destroy vast amounts of property and threaten lives. Oil spills like Ixtoc 1 create vast oil slicks, kill fish and birds, and blacken beaches. And the use of hydraulic mining—a powerful tool—during the Gold Rush utterly remade the areas where it was used, denuding hillsides, destroying forests, creating mudslides, and damaging rivers and streams beyond recognition. Obviously, the real culprit was not the water but how humans used that water or stood in the path of weather events by building their homes and business in harm's way. This is not to point the finger, but to demonstrate how awesome a natural force water can be under certain circumstances.

The book ends with a brief discussion of water and climate change. Several of the essays earlier in the book allude to that connection, but it is still important to end with a cautionary tale about the complex intersection of water with our changing world. Water is so integral to everything we do that we should not be amazed that our history is so intertwined with it. It is a precious resource, a mediator between humans and other aspects of our environment, and a mechanism of survival. It is not surprising that water has always played a central role in human culture and human society, but counting the ways is instructive.

Notes

Diamond, Jared (2005) *Collapse: How Societies Choose to Fail or Succeed* (New York: Penguin Books).

Gleick, Peter H. (2000) *The World's Water, 2000–2001: The Biennial Report on Freshwater Resources* (Washington, D.C.: Island Press).

Gleick, Peter H. (2006) *The World's Water, 2006–2007: The Biennial Report on Freshwater Resources* (Washington, D.C.: Island Press).

Gleick, Peter H. (2009) *The World's Water, 2008–2009: The Biennial Report on Freshwater Resources* (Washington, D.C.: Island Press).

Glennon, Robert (2002) *Water Follies: Groundwater Pumping and the Fate of America's Fresh Waters* (Washington, D.C.: Island Press).

McNeill, John (2000) *Something New Under the Sun: An Environmental History of the Twentieth-Century World* (New York: W.W. Norton).

Melosi, Martin V., ed. (2011) *Precious Commodity: Providing Water for America's Cities* (Pittsburgh, PA: University of Pittsburgh Press).

Outwater, Alice (1996) *Water: A Natural History* (New York: Basic Books).

Pritchard, Sara B., and Carl A. Zimring (2020) *Technology and the Environment in History* (Baltimore, D: Johns Hopkins University Press).

Reuss, Martin, and Stephen H. Cutcliffe, eds (2010) *The Illusory Boundary: Environment and Technology in History* (Charlottesville, VA: University of Virginia Press).

Rogers, Peter (1993) *America's Water: Federal Roles and Responsibilities* (Cambridge, MA: MIT Press).

Sedlak, David (2014) *Water 4.0: The Past, Present, and Future of the World's Most Vital Resource* (New Haven, CT: Yale University Press).

White, Richard (1995) *The Organic Machine: The Remaking of the Columbia River* (New York: Hill and Wang).

Part I

Indigenous Peoples Before Contact

In October 2021, sci-fi fans were ecstatic when Warner Bros. Pictures and HBO released the latest iteration of *Dune*. The film was based on the 1965 science fiction novel written by Frank Herbert. The cult classic focuses on Arrakis, the desert planet, which holds the only source of the life-prolonging drug mélange—or spice—and controlling it becomes the central cause for galactic conflict. More importance for our purposes is that water is the most important resource to make life possible on the planet. The native population, the Fremen, have learned how to survive on Arrakis, and as we learn, will ultimately thrive there despite attempts by hostile forces to subjugate them. The Fremen's knowledge and respect for water become their secret weapon.

Indigenous peoples in North America going back thousands of years demonstrated remarkable abilities to survive in a variety of settings and environments, not as inhospitable as Arrakis, but unhospitable enough. Water proved to be key to that survival, especially in the creative ways that they relied on and utilized water. Obviously, fresh water was necessary for life. But the versatility of fresh and salt water to serve other purposes was graphically demonstrated by the three cultures under study: the Hohokam of the American Southwest, the Aztecs of the Valley of Mexico, and the Inuit of the Arctic Zone.

The Hohokam constructed a vast irrigation system—the first large-scale system in the United States—beginning in about 850 CE along the convergence of the Gila, Verde, and Salt rivers in modern-day Arizona covering about 100,000 acres. Their elaborate technical system was the basis for an enduring society that shaped its environment on a scale unknown before. While some details of how the Hohokam culture collapsed are not known, some of their creative technical legacy was passed along to others and material remnants of their system imprinted in Arizona soil.

The Aztecs founded their capital city of Tenochtitlan on an island near the western swamps of Lake Texcoco in the Valley of Mexico in 1345 CE. Building their capital city on watery swampland may have seemed unrealistic, but the water barrier that it provided protected the Aztec from their enemies, and helped the city become a central base for regional power as well as for trade and farming. Without drayage animals or wheeled vehicles, the Aztecs

DOI: 10.4324/9781003041627-2

utilized canoes and rafts to move people and goods within the city and out into the periphery. They irrigated and employed chinampa (a raised field or garden) agricultural techniques to grow crops. We know about the conquest of the Aztecs by the Spanish, who employed tools and weapons not known or anticipated by the Aztecs when they built their defenses. Yet Tenochtitlan was a testament to a multi-faceted understanding of the value of water.

The Inuit carved out a successful lifestyle for about 4,000 years or more in the desolate and hostile Arctic Climate Zone in Greenland and in the northernmost parts of the Eurasian and North American continents. Like the Hohokam and the Aztec, water was an essential element of Inuit culture. In their case, however, they learned how to deal with water in a solid form as ice and snow. Their knowledge of sea ice, weather conditions, ice-covered terrain, and the habits of fish and wildlife led to successful adaptation to an environment that white explorers found too threatening. It remains to be seen to what degree the impacts of climate change will undermine Inuit adaptation to the evolving Arctic, or to what degree the Inuit can help us find sustainable ways to live in a world facing threatening environmental conditions. Inuit Tapiriit Kanatami'S (the national representative organization for the 65,000 Inuit in Canada) *National Inuit Climate Change Strategy* (2019) provides information on the vision of Inuit leaders for combatting climate change. See also Zoltan Grossman and Alan Parker's edited volume, *Asserting Native Resilience: Pacific Rim Indigenous Natives Face the Climate Crisis* (2012) for examples of how other indigenous peoples are confronting the climate crisis.

Each story is fascinating. Each story is unique. The common element is the way these cultures adapted to their surroundings—desert, high plains, and arctic—utilizing water as a central component in their growth and survival. The stories also are suggestive of a variety of themes for further exploration. How do societies respond to aridity and drought, and with what tools or actions? Georgina Endfield takes on that issue in *Climate and Society in Colonial Mexico* (2008). In all the water episodes in the book, the relationship between water and culture is vastly important. A good start is Terje Tvedt's nine volume *A History of Water* (2006–2016). Of course, looking at how other indigenous people acquired their water and how they used it makes for excellent comparisons among peoples. And, as colonialism came to the Americas, questions of indigenous water rights are critical.

1 The Hohokam: "The Canal Builders" of the American Southwest

Water meant survival for the Hohokam, and survival rested upon the establishment of an agricultural society dependent on irrigation on a large scale. Their canal system mainly in the Gila and Salt River Valleys covered more than 100,000 acres of mostly desert land in what is now primarily southern Arizona. The Hohokam set a standard for constructing large and complex irrigation networks using preindustrial technology with future implications for farming in the arid American West.

The sprawling city of Phoenix in Arizona's Valley of the Sun sits atop numerous sites of the Hohokam civilization that appeared along the convergence of the Gila, Verde, and Salt rivers from about 200 CE to 1450 CE. The name *Hohokam* likely came from the Pima meaning "those who vanished" given their ambiguous disappearance after 1450. Archeologists usually divide Hohokam history into four periods: Pioneer (200–775), Colonial (775–975), Sedentary (975–1150), and Classic (1150–1450). According to Marc Reisner in *Cadillac Desert: The American West and Its Disappearing Water*, the Hohokam established a civilization that "rivaled the Aztec, Inca, and Maya farther south" (Reisner, 1993, 256).

At first, the early desert cultivators planted crops on the fertile floodplain, and later descendants began diverting water into their fields through short and shallow canals. We now know that the Hohokam built a vast canal network and irrigated as much as 110,000 acres of desert land serving as many as 50,000 people. A Phoenix city engineer who mapped the irrigation system in the 1920s estimated that it was the largest tract of irrigated land in prehistoric North and South America. In addition, the Salt River Valley was the most populous and possibly the most agriculturally productive area in the American Southwest before 1500.

For our purposes, the story of the Hohokam is a tale of determined and talented people prevailing in an extremely hostile environment through its innovative capture of water. A difficult question to answer is whether they were attempting to subdue the natural world for survival, or trying to find a way to adapt to it?

Hohokam society existed within the Phoenix Basin in the extraordinarily hot and dry Sonoran Desert, which extends from southwestern Arizona to

DOI: 10.4324/9781003041627-3

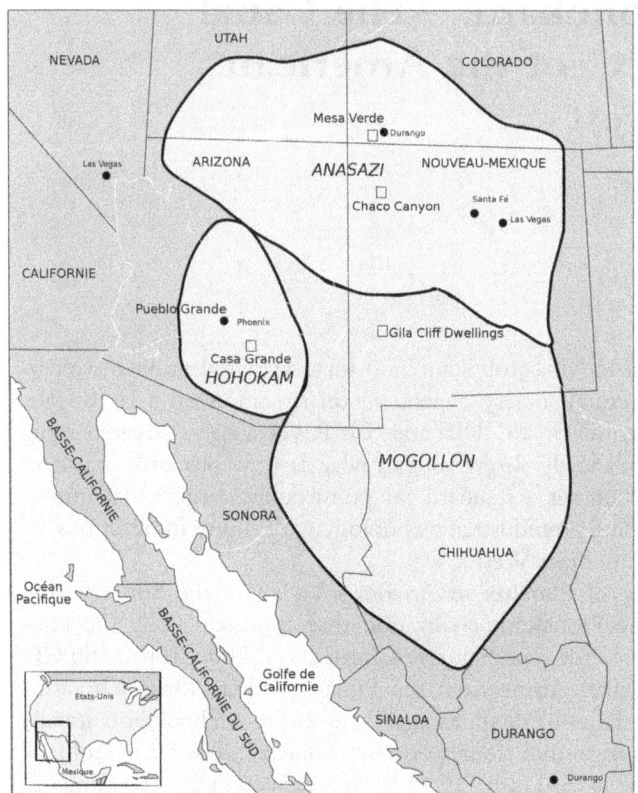

Figure 1.1 Map of the Hohokam Area ca. 1350. Unknown author. Creative Commons Attribution-Share Alike 3.0. Wikimedia.

southeastern California, and into northwestern Mexico. It is over 192,625 square miles in size. In the area in which the Hohokam's villages grew, the mean annual precipitation ranged between 6 and 15 inches—most often favoring the lower end of the scale. There were two distinct rainy seasons, one in the winter due to storms from the Pacific Ocean, and another in the summer with thunderstorms driven by subtropical forces from the south. Winter rains contributed extensive runoff and destructive flooding along the Salt River. Summer flooding tended to be smaller and less destructive. Temperatures in the summer could exceed 105 degrees Fahrenheit. The topography of this region consisted mainly of low-level slopes. At the basin floor the elevation is less than 985 feet, and about 3,937 feet at the summit of the McDowell Mountains to the northeast of the basin floor.

The Hohokam canal system was adapted to the weather conditions and to the desert terrain. In that area the Gila and Salt rivers received their water from highland watersheds. The Sonora Desert, however, provided many resources to sustain life. Hohokam communities could accumulate surpluses of storable food such as fruit from cacti and seeds from trees and shrubs. Although they were not a hunter culture, the Hohokam had access to deer, bighorn sheep, small game such as rabbits, fish, and mussels. Less abundant was wood for fires and building purposes, but ironwood, paloverde, mesquite, and cottonwood could be found along drainage areas. Ultimately, the Hohokam not only emerged as premier agriculturalists but supplemented farm production with foraging and short- and long-distance trade.

Over time, many Hohokam farming communities grew along the rivers. What archeologists call "sedentary villages" (places where subsistence economies were established) predate the Hohokam, as did farming along the desert rivers going back as early as 1200 BCE. But the scale of the Hohokam's village development and its farming was much greater. Early settlements in the Pioneer period concentrated along the major rivers but were scattered with individually built structures made of wood, clay, and brush and sat atop shallow pits. In the Colonial period, settlements were established on a more regional scale. They could be in upland areas to the south and west of current-day Tucson and Phoenix, and between Phoenix and Flagstaff. Villages in general varied greatly in size with some covering 15 square miles and containing more than 2,500 irrigated acres.

Snaketown, which was excavated in 1934–1935, was located about 30 miles southeast of Phoenix and may have had as many as 2,000 inhabitants. Snaketown apparently got its name because the heaps of garbage collected by the Hohokam became nesting places for rodents, which attracted rattlesnakes and other serpents looking for a meal. Snaketown was considered the cultural capital of the Hohokam and a production center. It is significant because it provided evidence of broad Mesoamerican (a cultural area including modern-day Mexico, Guatemala, Honduras, Belize, El Salvador, Nicaragua, and Costa Rica) influence in the region, such as the cultivation of maize and squash, the use of ballcourts for playing a Mesoamerican ballgame, and the presence of some common trade items, such as copper bells, pyrite mirrors, obsidian, and turquoise. Evidence of artifacts, such as red-on-buff pottery made in Snaketown have been found in surrounding communities. Indeed, the Hohokam may have originated in Mesoamerica and migrated northward into the American Southwest.

The central feature of the Hohokam's contributions to modernity was its hydraulic system, which was an engineering marvel matching that of the ancient Roman aqueducts. In the New World, it was the most massive irrigation system north of Peru. Complementing the Hohokam engineering prowess was their farming skills, including domesticating one or more varieties of maize (new to the Southwest), as well as growing cotton, beans, and squash.

Canals were constructed between 850 and 1450, reaching their greatest extent in the Classic period. About 150 miles of canals in all were in the Salt River Valley. The decision to irrigate the desert environment may have been one of necessity, especially when the Hohokam shifted to a sedentary, agricultural life with a larger population requiring more water to produce more crops. The Hohokam, however, practiced both canal irrigation expanding within the core communities and dry farming in some areas along the periphery of those communities.

In the spring, the Hohokam concentrated on maintaining and repairing their irrigation systems, while in the summer they spent time gathering wild plants and developing thousands of acres of farming fields. The canals often extended 16–22 miles in length. One canal unearthed in Mesa was 15 feet deep and 45 feet wide.

Overall, the Gila and Salt River valleys were ideal settings for a massive irrigation system because of their relatively flat topography. The main canals operated by gravity, achieving a gradient (drop) of only a few feet per mile. They were tapered to allow the water to flow at a constant rate. Such an approach lessened channel erosion from fast-moving water and from silting that occurred when water slows down. The canals transported water laterally as well as downstream for miles beyond the source, which increased coverage dramatically. The builders faced different problems on different rivers. Patterns of water discharge and the form of the channels depended on the amount of water available, sediment load, and the kind of streambed they encountered. In the Salt River Valley, for example, the topography allowed for the extending of canals as far as 8 miles from the river and in a variety of directions; along the Gila's main channel the canals had to parallel the river in most cases.

The canals themselves consisted of a variety of components. In places where the canal met the river, a weir was likely constructed. (A weir is a dam that reaches into part of a river). It raised the level of the water in the river and directed it into the canal. Inside the canal proper there was a headgate (water control gate) to regulate the amount of water entering the canal. The main canals were large at the juncture with the river, reduced in size as the water traveled to the fields. Distribution canals carried water from the main canal to the fields and were a way of controlling the water level in the canal by utilizing a variety of control features, such as diversion gates or control gates. Water control gates (tapons) were often located inside the main and distribution canals. The control mechanisms allowed the Hohokam to create a highly complex irrigation system.

Rather than relying on some sort of mechanized construction technology, the Hohokam depended on vast numbers of workers to build the canals—and without wheeled vehicles. The soil was loosened by hand with "stone hoes" (large wedge-shaped pieces of stone) and wooden digging sticks and removed in large baskets. Reconstruction of some of the canals suggests that more than 28 million cubic feet of dirt may have been removed to produce a major canal. "Leveling frames"—used in various preindustrial agrarian cultures—were used to establish gradients of the canals. Water may have been carried along the canal during construction to loosen the soil.

Canals and their component parts could not last forever once constructed. Flooding in particular, but regular usage, in general, could damage or destroy

Figure 1.2 A Restored Hohokam Canal in Arizona. MWyattB. Creative Commons
Attribution-Share Alike 4.0. Wikimedia.

the canals, requiring repairs or rebuilding. One estimate suggested that individual canals seldom exceeded 50–100 years in life span, which was significant given the duration of Hohokam society in the Sonoran Desert. Also,
canals had to be dredged periodically as to not obstruct water flow.

One of the more difficult questions to answer is: To what degree did the
building and maintenance of the canal system involve a central authority?
This is particularly important because such a vast and complex undertaking
required a well-organized effort, and strong leadership was necessary to
resolve conflicts. There is little doubt that the system needed some degree
of interdependency among the increasing number of households in the
many villages. The extent to which this interdependency demanded specialized and hierarchical structures beyond kinship relations is not altogether clear. The Hohokam, however, exhibited some type of inequality in
power and status.

The precise role of elites in coordinating labor for the construction of the
canals and public monuments has not been resolved by archeologists. It is
likely that those who participated in the canal construction had the pick of the
best land. Individuals or families with early access to the premium irrigated
land would acquire wealth and status. Those who arrived later would have to
occupy the remaining acreage.

Rights to irrigation water belonged to the community as a whole. Water
was allocated to each household according to the amount of land cultivated.

Households thus shared water rights. On a broader scale, Hohokam society was made up of "irrigation communities" with its own leadership to organize labor for main canal construction and maintenance. Small, local groups of farmers could organize for the construction and maintenance of the secondary branch and distribution canals.

The canal networks connected groups of interdependent villages, whose households shared the construction, maintenance, and even management of the canals. It is quite likely that competition among Hohokam settlements or blocks of settlements operated instead of some central authority to control the canal system. Especially in areas of intense irrigation, it was common to find inter-village coordination. Settlements at canal headgates had direct control over allocation of water. It is possible, but little evidence is available, that the threat of force by settlements at the ends of the canals could compel those on the headgates to cooperate. But settlements sharing a canal had good reason to cooperate with each other in the long term in order to keep the water flowing.

Some researchers believe that "platform mounds" constructed by the Hohokam had a relationship to the organization and operation of the canal systems. A platform mound was an earthwork meant to support a structure or a particular activity. Administrative sites, containing one or more platform mounds could be found at the headgates of the major canals (such as at Pueblo Grande, Mesa Grande, and Tres Pueblos). The flow of water could be controlled from these sites, and labor necessary for annual repairs could be organized. Other platform mounds were located at regular intervals (usually 3 miles) along the canals.

Some archeologists surmised that Hohokam elites lived on the platform mounds, but there is little direct evidence. Large residential compounds were built closest to the mounds, with smaller compounds beyond them. The non-random arrangement of the residences suggests a degree of social ranking in relation to the mounds.

In addition, platform mounds may have had a religious function. This was based on the fact that rooms in the complexes tended to be separated from each other, which meant segregation of activities (even secrecy). This feature often was found in religious architecture. Some scholars speculate, therefore, that Hohokam religion may have played an important role in organizing the canal system.

While there is little available evidence of the state of gender roles, women primarily were responsible for household duties among the Hohokam. One archeological site identified menstrual huts where menstruating Hohokam women were isolated, perhaps because they were regarded as dangerous to the community. Egalitarianism, if it existed, did not extend to women. One anthropology study dealing with the Pre-Classic to Classis period, however, suggested "that some women occupied domestic spaces and were buried on top of platform mounds indicates that these women were accorded higher honor than both men and women occupying compounds and buried off mounds, in most instances probably because of ties established through birth and marriage." Also, for the most part, "prestige derived from different realms for males and females" (Crown and Fish, 1996, 812).

Approximately between 1150 and 1450, the relative stability of Hohokam society was undermined because of a change in climate. Unstable environmental conditions dominated by droughts—causing water shortages and flooding—altered river channels and damaged or destroyed canals. The Great Drought of 1276–1799 appears to be noteworthy. Between 1286 and 1316, there were four major droughts and five major floods. Therefore, the irrigation system could not always produce the food surplus to support the once dynamic communities. Climatic fluctuations damaged the alignment of canals. Also, according to environmental historian Donald Worster, "…the Hohokam overreached themselves. Intensive irrigation has everywhere led to increasing concentrations of salts in the topsoil, poisoning the farmers' fields. That nemesis came to the Hohokam too, and they were forced one day to abandon their agriculture completely, leaving behind them whited fields and dust-drifted canals" (Worster, 1985, 34).

Between 1350 and 1450, tribe members abandoned most of their settlements. Hohokam society thus collapsed as households moved away to farm on a smaller scale or to settle with kin. But it should be remembered that collapse was taking place over much of the Southwest, not simply among the Hohokam.

After the Hohokam left, the Pima and Papago (Tohono O'odham)—thought to be direct descendants—occupied some of their lands and borrowed some of their water technology. The Papago and other Sonoran groups utilized "floodplain irrigation," limited to a few river edges and arroyos. Some of the Hohokam's abandoned canals were sold to Gold Rush prospectors in the Phoenix area. In 1867, former soldier Jack Swilling established the Swilling Irrigating and Canal Company. In 1871 settlers selected a townsite—Phoenix—to be built on top of some of the Hohokam's largest settlements. North of Arizona in what became Utah, Mormon settlers constructed an irrigation system in the mid-nineteenth century in the Great Salt Lake Valley. Some claimed that the Mormons were the first to use large-scale irrigation in the American West. They were wrong.

The Hohokam were an amazingly enduring society. They built an elaborate technical system with hand labor and primitive tools on a scale few could imagine—especially in the inhospitable Sonora Desert. Water was the essential element for life, for sustaining their people, for providing agricultural riches. To accomplish their goals, the Hohokam left a grand environmental footprint that changed the rivers and valleys of Arizona, not only for their time but for the future as well.

Their canal system became a timeless imprint demonstrating that humans were capable of extraordinary things. For some, they had conquered nature and won. For others, despite the achievements, the scale of the Hohokam's system probably was too big to manage and was vulnerable to the pressures of climate and other forces that they could not ultimately subdue.

As anthropologist Suzanne K. Fish stated, "The Hohokam are noted for the most enduring settlements of the prehistoric Southwest" (Lentz, 2000 274). Their settlements were enduring because they gave the highest priority to accessing fresh water in an environment bent on withholding it.

Notes and Further Reading

Abbott, David R. (2000) *Ceramics and Community Organization among the Hohokam* (Tucson, AZ: University of Arizona Press).

Bayman, James M. (2001) "The Hohokam of Southwest North America" *Journal of World Prehistory* 15 (September): 257–311.

Crown, Patricia, and Suzanne K. Fish (1996) "Gender and Status in the Hohokam Pre-Classic to Classic Transition, *American Anthropologist* 98 (December): 803–817.

Haury, Emil W. (1976) *The Hohokam: Desert Farmers & Craftsmen* (Tucson, AZ: University of Arizona Press).

Lentz, David L., ed (2000) *Imperfect Balance: Landscape Transformations in the Precolumbian Americas* (New York: Columbia University Press).

Reisner, Marc (1993) *Cadillac Desert: The American West and Its Disappearing Water* (New York: Penguin, rev. ed).

Rice, Glen (1998) "War and Water: An Ecological Perspective on Hohokam Irrigation," *Kiva* 63, 263–301.

Seelye, James E., Jr (2011) "Southwest Indians," in Kathleen A. Brosnan, ed, *Encyclopedia of American Environmental History*, v. III (New York: Facts on File), 758–760.

Worster, Donald (1985) *Rivers of Empire: Water, Aridity & Growth of the American West* (New York: Pantheon Books)

2 The Aztecs and the Founding of Tenochtitlan

The primary function of water among the Hohokam was providing a livelihood based on large-scale farming and for drinking, cooking, and cleaning. In the Aztec world water would have a similar function, but also—in and around its capital city of Tenochtitlan—it would constitute a central means for defense of the city and also serve as the major mode of transportation for access to other resources and for trade.

In Aztec religion, Chalchiuhtlicue was the water goddess and wife of the rain god Tlaloc. Rain was an essential part of life, especially in arid regions of Mexico. But Chalchiuhtlicue also symbolized the purity and value of spring, river, and lake water necessary to irrigate the fields. These deities like those in other cultures accentuated water as the sustainer of life. What made the Aztecs special was their ability to find ways to think about water in multiple, and quite ingenious, ways.

Tenochtitlan (Mexico-Tenochtitlan) was the capital city and religious center of the Aztecs. It was founded in 1345 CE on an island near the western swamps of Lake Texcoco in central Mexico. Texcoco was one of five interconnecting waterbodies on the floor of the Valley of Mexico. At the time, Tenochtitlan was among the largest cities in the world. In 1521 CE, Spanish forces led by Hernan Cortes destroyed the city, which led to the collapse of the Aztec Empire. Mexico City arose on the ruins of Tenochtitlan.

The Aztec civilization thrived in the latter stages of the Postclassic Period (about 900–1519) in Mesoamerica. According to Latin American historian Jonathan C. Brown, the period was characterized "by the development of aggressive imperial states and large-scale warfare. Mesoamerica was "a mosaic of languages and ethnic diversity," with as many as 260 major languages between the northern border of Mexico and the southern border of Guatemala. Late in the period the Toltec and Aztec people spread the Nahuatl language throughout the region. Brown added that the Postclassic empires were "much more warlike and predatory than the city-states of the previous period" (Brown, 2000, 22). This was the world in which the Aztecs emerged.

The migration that led the Aztecs from Aztlan to Tenochtitlan is steeped in myth, legend, and history. But the migration, which took many years,

DOI: 10.4324/9781003041627-4

also is a story of "unrelenting struggle, rivalries, conflict, suffering, and eventual triumph," according to a historian of religion and anthropologist David Carrasco (Carrasco, 2012, 17). Of the various migrant groups linked to the Aztec culture, most are known about the *Mexica*, the group that would come to dominate it and found the great city of Tenochtitlan. The story about the migration of the Mexica (composed of two migrant groups) to the highland Valley of Mexico is a composite of stories—some real and some imagined—that became the orthodox version later found in Mexican schoolbooks.

The Central Valley of Mexico and adjacent valleys to the north, east, and west are part of the Central Mexican Plateau comprising several ecological zones. It is arid to semiarid, occupying much of northern and central Mexico averaging almost 6,000 feet above sea level. The Valley of Mexico itself varies in altitude between 6,800 and 7,900 feet. Volcanic activity, glaciation, and erosion created an array of soils. The bottom of the valley initially was covered with extensive lakes—saline and fresh. The rivers, however, were not well suited for irrigation. The mountainous terrain created by volcanic activity blocked the course of rivers to only short distances. Mexico does not have any area where inland waterways extend for hundreds of miles. Thus early peoples in the area could not have developed water systems to feed agricultural plots; not even the abundant lakes were suitable for large-scale irrigation.

Rainfall in the Central Valley ranges between about 20 and 31 inches and is limited mostly to the months of May through October. Temperatures are substantially cooler than other zones, and frost can be a problem for farmers from October to March. The growing season, therefore, is shorter than adjacent regions. Central Mexico, eventual home to the Aztecs, is mountainous with steep wooded slopes. As recently as 1,000 BCE there were less than 10,000 inhabitants in the valley.

Some areas had adequate rainfall for farming and in others not enough to support the growing of maize. Early farming communities (1000–500 BCE) were most concentrated along the southern slopes of the valley where rainfall was most plentiful. They cleared hillside land and enclosed it in terraces to prevent erosion, thus allowing them to acquire water flow down the mountain slopes.

In the ancient Teotihuacan city-state (flourishing between 150 and 700 CE), located about twenty-five miles north of modern Mexico City, farmers used water from underground springs. Small groups dug crude irrigation ditches to divert spring water to their fields. Such practices led to development of more elaborate irrigation and higher production of crops. By the first century CE, a medium-sized urban settlement existed at Teotihuacan; two centuries later the population reached about 20,000, and by about 500 it exceeded 100,000. Their agricultural system also spread elsewhere. Unfortunately, by about 700, Teotihuacan faced raids from invading groups and collapsed by about 750 (Oaxacan and Mayan civilizations also collapsed about this time).

Three centuries later, the population of the Valley of Mexico had dropped to about 150,000 people. Irrigation technology was not forgotten, but there were no great city-state communities to produce a labor force large enough to construct and maintain a sizable system. Bands of migrants continued to enter the valley. In Tula, the Toltecs built a community patterned after the architecture of Teotihuacan, as well as borrowing its irrigation technology. The Toltecs soon spread beyond Tula, controlling large parts of the Valley of Mexico. But their reign lasted only about 200 years, possibly because of drought and crop failures and eventual attacks from new groups in the twelfth century.

As early as the eleventh or twelfth centuries, the Chichimecs (nomadic hunting tribes) departed from their homeland somewhere in the north called Aztlan, "place of cranes (or herons)," from which the generic name *Aztec* is derived. It was located anywhere from just north of the Valley of Mexico to southwestern United States. Some say that it was near one of the lakes in Michoacán, Jalisco, or Guanajuato; others suggest it was near a coastal lagoon on the Pacific coast of Nayarit. Motecuhzoma I, who ruled Tenochtitlan from 1440 to 1469, sent an expedition to find it. The party traveled north beyond the ruins of Tula, returning only with a story of a mythic place, not a specific location. They may have been correct that Aztlan simply was imaginary.

There were at least two waves of migration from the north into the Valley of Mexico with the second one around 1250. The last to arrive were the Mexica. Existing records have more about this group than earlier migrants, and it became the generic name for those who founded Tenochtitlan and the name of the migrant settlers known by the Spaniards.

The Mexica settled in various locations along their journey for as much as twenty years at a time. They farmed and built elementary temples and ballcourts at some locations, but always moved on because of a deified ancestral hero Huitzilopochtli's vision of destiny or because they were rebuffed by others. During the migration, priests carried a huge idol of the patron god Huitzilopochtli, who apparently whispered directions along the trek.

Eventually, tribal quarrels broke out, splintering the group. The tattered main body of castaways moved on to Tula before continuing to the Valley of Mexico, eventually arriving in a largely impoverished area called Chapultepec. There they faced a hostile reception and were driven out. Copil, the chieftain who led the attack against the Mexica and the son of Huitzilopochtli's sister, was killed. Legend says that the great war god ordered that Copil's heart be thrown onto an island in Lake Texcoco, and there the Mexica should build their new home.

The Mexica returned to Chapultepec but were defeated again, with refugees escaping to the countryside while their leader was taken to Culhuacan to be sacrificed. At this point, they were forced to give up farming and return to hunting and gathering for many years. Eventually, the Mexica refugees made their way to Culhuacan to plead for protection of its rulers. The group was granted land at Tizaapan—a desolate, snake-infested lava flow in the southwestern part of the valley—and served as military allies of the city of Culhuacan. The Mexica began to regard themselves as "Culhua-Mexica," also believing they were descendants of the deeply admired Toltecs. While the Culhua aristocrats viewed the Mexica as barbaric and inferior, they were beholden to them because of the central role they played in defeating neighboring Xochimilco. With long memories of their virtual subjugation, however, the Mexica turned on the Culhua but were driven into the swamps of Lake Texcoco.

Utilizing canoes and rafts, they headed across the lake to an uninhabited marshy island, Tenochtitlan, in 1345. Once there, one of Huizilopochtli's priests claimed to have a vision in which the hero god repeated that Copil's heart had been thrown on the island and that on the spot there would be a nopal cactus upon which an eagle was perched with a snake in its mouth. Finding the eagle on a cactus in the reeds, they constructed a crude platform with a hut as a shrine to Huitzilopochtli. On that site, the Great Pyramid of Tenochtitlan eventually would be built. The marshy site became the location of their permanent home. The name, *Tenochtitlan*, was derived from *tetl* (rock), *nochtli* (prickly-pear cactus), and *tlan* (a locative suffix).

The new city was located between three important settlements: Culhuacan to the south, Atzcapotzalco to the west, and several smaller towns along the eastern side of Lake Texcoco. Like earlier city-states in central Mexico, Tenochtitlan had a formally planned city center, but un-planned residential areas. Borrowing principles and ideas from the ruins of older civilizations such as Teotihuacan and Tula, the Mexica nevertheless wanted to produce something grander and more distinctive. The grid layout and architectural and sculpture style copied past practices, but the city's sacred religious precinct was walled off from the rest of the structures, unlike its predecessors. The influence of religion was very powerful in the overall design of Tenochtitlan.

At the center of Tenochtitlan was the walled religious and ceremonial precinct—the Sacred Precinct—containing as many as 78 separate edifices. It was dominated by a main temple (Templo Mayor) upon which sat two shrines—one to the patron god Huitzilopochtli and the other to the very important supreme god of rain, Tlaloc. Additional structures included temples to various lesser gods, a ballcourt, priests' quarters, schools for training nobles for the priesthood, administrative buildings, and three entrances leading out to four main residential quarters. There also were adjacent opulent palaces with gardens, aviaries, and zoos. Common people (including craft specialists, la-borers, and farmers) lived at a distance from the Sacred Precinct and were organized into neighborhoods or barrios with their own temples and markets. Commoners lived in mudbrick houses.

Ritual sacrifice was conducted primarily at the two sacrificial stones at the Templo Mayor in the Sacred Precinct (and elsewhere in the Aztec Empire), sometimes on a grand scale. There is considerable controversy about human sacrifice in the ancient world (especially with the Spanish exaggerating the

Figure 2.1 El templo mayor en Tenochtitlan.png. Donde ocurre la matanza del templo mayor. Ckn8u. Creative Commons Attribution-Share Alike 4.0. 5 May 2016.

numbers), but it was practiced to some extent by several cultures including the Greeks, Hebrews, Mayans, and Incas. Behind the practice were strong religious overtones. The Aztecs believed that the gods sacrificed themselves to create the sun and the world, offering their own blood to make humans. As a result, humans owed their allegiance to the gods, especially through offerings of human blood. Among other things, such sacrifices could enhance agricultural or human fertility. This could take one of two forms: autosacrifice or human sacrifice.

Aside from viewing sacrifice exclusively from the vantage point of ritual for ritual's sake, some scholars have argued that famine could prompt large-scale ceremonies of this grim activity. Sacrificing to the war god Huitzilopochtli also could suggest a way to frustrate vassal states from challenging Aztec authority.

Most of the major construction in Tenochtitlan took place in the 1470s, during the reign of four Aztec kings. The city was built primarily on land reclaimed from the swampland on the southern portion of the island and in shallow Lake Texcoco itself (with some land taken by force from neighboring city-states). The water in the immediate area of the future Tenochtitlan was brackish, and thus too salty for potable use or irrigation. The area, however, did possess an abundance of fish, frogs, turtles, birds, and even algae, and had underground springs for drinking water. But the greatest virtue of the city's location was its immense strategic and defense potential, offering an excellent barrier between the Aztec leaders and tribal rivals on the mainland.

At its peak, Tenochtitlan covered an area of only about 4.6–5.4 square miles and was laid on a grid pattern (inspired by Teotihuacan) with numerous canals dispersed throughout. There were four main thoroughfares dissecting the city, with adjacent narrow streets and canals. Specifically, there were six major canals running east and west through the city, and two major canals running north and south. Since there were no wheeled carts or wagons, nor any drayage animals, goods were transported by thousands of porters, small boats, or canoes. The city was connected to the western shore of the lake and nearby countryside by three raised causeways running north, east, and west. The causeways were linked to nine smaller urban settlements on the mainland and to farming areas. In the gaps were removal bridges allowing boats and canoes to pass. The bridges could be taken down in case of an attack on the city. Tenochtitlan also had a stone aqueduct (maybe two) bringing in fresh water from springs near Chapultepec Hill.

The neighboring city of Tlatelolco on the north side of the island had a huge market and was a center of commerce and artisanry. While the Mexica in Tenochtitlan were perceived as a warrior and bureaucratic group, Tlatelolco attracted merchants, artisans, and farmers from various subject peoples. Originally some of the migrant clans broke off from the main body that first inhabited Tenochtitlan and settled in the island-lagoon of Tlatelolco. The community became a major rival of Tenochtitlan because of its market. In 1473 the Mexica emperor Axayacatl conquered Tlatelolco and annexed it into the main city. (At some point it may have been linked by landfill to Tenochtitlan as well.)

Tenochtitlan became the center of a vast tributary and trading center. Markets were set up in Tenochtitlan and Tlatelolco in what became the start of a more elaborate trading network. Commerce with other towns in the valley occurred using boats and canoes. The Mexica obtained construction materials through the market system, utilizing fish, frogs, ducks, and algae in exchange for such materials.

The Mexica exploited the full potential of lake, river, and groundwater around them for defense, potable use, transportation, commerce, and waste disposal. Leaders also began efforts to construct a *chinampa* system, which they had learned from the Culhua, on the freshwater part of the island in the fourteenth and fifteenth centuries. The chinampa—raised fields or gardens made of roots, branches, bushes, trees, and brush covered with soil from the lake bottoms and shaped into long rectangular islets—were built on reclaimed land and firmly connected by poles or cypress trees to the shore or the lakebed. (Some chinampas had been seized from neighbors.) The beginnings of chinampa agriculture in the Valley of Mexico are unclear. There is some evidence of it around Teotihuacan in the first millennium CE and possibly earlier.

Using a system of dikes and canals, the builders kept salty water from Lake Texcoco separate from fresh waters of Lake Chalco and Xochimilco, which circulated among the chinampa gardens. The porous soil and flowing water created a fertile environment that attracted birds, fish, algae, and insects. Eventually, the cultivated fields on the outskirts of Tenochtitlan formed a green ring around the inner city and provided year-round agricultural production of vegetables, cereals, and fruit. The largest extension of chinampas took place on the southern part of Lake Chalco. The chinampas supplemented (and in some cases surpassed) land-based cultivation in the area.

In addition, by the time of the Spanish Conquest almost all available sources of water in central Mexico, such as large segments of the Cuauhtitlan River in the northwest part of the valley, had been utilized for irrigation. The Aztecs were not the first to employ irrigation in the area, but their systems tended to be larger and possibly more sophisticated. The longest canals were constructed of stone and channels were lined with plaster. Aqueducts brought water to Tenochtitlan's central plaza, and much of it was split off into clay pipes to temples, to residences in the sacred precinct, and to the palaces and villas of the dignitaries. Additional pipes carried water into individual rooms and pools in the residences of the nobles and then was drained away into the canals.

The commercial and agricultural prowess and the increasing consolidation of power of the burgeoning Aztec Empire contributed to its astonishing population growth. In the 1450s, the population was about 70,000 in Tenochtitlan; by 1519 a little more than one million people were living in the Valley of Mexico, and possibly another two or three million Aztecs lived in the surrounding valleys of central Mexico. Their water policies helped to sustain the growing population.

Once a weak, people were forced to pay tribute to powerful city-states, the Mexica of Tenochtitlan formed the *Triple Alliance* (1428) with Texcoco (to the northeast) and Tlacopan (on the western mainland) initially as a war pact against the Tepanecas. In 1433, this alliance conquered the powerful city-state of Atzcapotzalco—to which it had been paying tribute (in what amounted to the final century of Aztec civilization). The Triple Alliance would rule the largest empire ever built in ancient Mesoamerica. Ultimately Tenochtitlan's enhanced military power allowed the Mexica to dissolve the Triple Alliance and subordinate Texcoco and Tlacopan, and extend the empire outside of the Valley of Mexico. It now ruled the empire largely uncontested but was not immune to chronic warfare.

The city-state of Tenochtitlan was the greatest symbol of the power of the Aztec Empire. Awash in tributes from neighboring peoples, it was the center of great wealth and an ever-rising nobility led by a series of powerful emperors. The cult of Huitzilopochtli was at the center of power among the empire's leaders.

The Aztec Empire itself represented a wide variety of peoples, cultures, and languages, where local elites ruled their subjects and collected tribute for the emperor. The society's strict class system, however, dictated the size of one's house and even the clothes one wore, which reflected the top-down nature of

Tenochtitlan. In this system, gender roles were strictly defined with women restricted primarily to domestic activities. Slavery also existed but represented only about five percent of the total population.

As Brown stated, however, "Among the Mexica and other Nahuatl-speaking peoples of the Valley of Mexico, the *calpolli* was the basic unit of social organization; it lent a degree of democratic decision making and participation to an autocratic, imperial political system in which the emperor was practically deified" (Brown, 2000, 32). A *calpolli* –or family clan–was a group tied by birth and kinship (through a common deified ancestor) living in a specific location, where land was held communally (although they believed that a local god gave them permission to use it). State leaders acknowledged a calpolli's connection to its land, which essentially bound the members to work it. The calpolli also organized participation in military engagements. The leader of the calpolli was elected but usually came from a local noble family. The term calpolli also came to denote neighborhoods in the city.

The destruction of Tenochtitlan by Cortes and his Tlaxcallan allies in 1521 devastated the grandeur of the Aztec culture. For our purposes, however, one of the most significant aspects of that culture continuing to this day was its embrace of water as fundamental to power, growth, and survival in terms of strength and defense, agricultural abundance, vigorous commerce, and mobility. As professor of Art History, Barbara E. Mundy, stated, "The unusual watery environment of the [Valley of Mexico] within which Tenochtitlan was built and the means of controlling it have been a source of wonder and puzzlement in all its histories..." (Mundy, 2015, 27). And also a central ingredient in sustaining Aztec power.

German-American historian and political scientist Karl A. Wittfogel referred to civilizations that were dependent on large-scale waterworks for irrigation and flood control as "hydraulic societies." Such civilizations, he argued, existed in ancient Egypt, Mesopotamia, India, China, Peru, and Mexico. Wittfogel linked water and social control. By transforming their environments (in this case through irrigation and related technologies), humans altered the basic conditions of their lives. For example, where irrigation required centralized control, governments monopolized political power and dominated the economy, controlled labor, and might even identify with the dominant religion. This would lead to despotism.

Other scholars have refuted Wittfogel's claim or modified it. But the claim does raise some interesting questions about the Aztec Empire, and the degree to which control of water had broad political ramifications, or if the centralizing tendencies of the Aztecs influenced how they used water in its various ways. Another point: What about the case of the Hohokam irrigation system? Does its history fit into the Wittfogel thesis? This is a question worth pondering. But there is little doubt that one cannot think about the rise of the Aztecs and Tenochtitlan without thinking about water.

Notes and Further Reading

Brown, Jonathan C. (2000) *Latin America: A Social History of the Colonial Period* (New York: Harcourt College Publishers).

Carrasco, David (2012) *The Aztecs: A Very Short Introduction* (New York: Oxford University Press).

Kandell, Jonathan (1988) *La Capital: The Biography of Mexico City* (New York: Random House).

Mundy, Barbara E. (2015) *The Death of Aztec Tenochtitlan, The Life of Mexico City* (Austin, TX: University of Texas Press).

Smith, Michael E. (2012) *The Aztecs* (West Sussex, UK: Wiley-Blackwell, 3rd ed.).

Townsend, Richard F. (2009) *The Aztecs* (London: Thames & Hudson, 3rd ed.).

3 The Inuit, Sea Ice, and Snow

Water comes in fresh and salty varieties and is found in many places. Depending on it for drinking, cooking, cleaning, defense, energy, and transportation, is more difficult when that water comes as ice and snow. It is amazing that the Inuit, among other peoples in the frozen North, have adapted so successfully to a hostile environment dominated by frozen water.

What we know of the Inuit's precontact past demonstrates an adaptation to a setting where water was no less important to the Inuit than to the Aztecs in central Mexico and the Hohokam in the American Southwest. Yet the form and function of water as ice and snow offered a very different set of challenges. The ability to adapt—and to adapt successfully—to one's environment is a common thread for all three cultures but played out very differently in the Inuit's world.

For more than 4,000 years, culturally similar people known as the Inuit inhabited Arctic regions spanning 12,000 miles from the Chukchi Peninsula in Siberia, and then east across Alaska and Canada, to the southeastern coast of Greenland. Although the Inuit are one of the most widely dispersed groups on the planet, the majority now reside in Canada's northern regions. They refer to their homeland as *Inuit Nunangat* (Inuvialuit in the northern parts of the Northwest territory and the Yukon; Nunavut and Nunavik in northern Quebec; and Nunatsiavut in northern Labrador). However, they share the polar region with several other indigenous cultures.

The Inuit are related to Mongoloid peoples of eastern Asia. *Inuit* means "the people" (One person is called *Inuk*.). The name *Eskimo* has long been used by non-Inuit people to describe them in popular culture, various writings, and among the general public. It is possibly of Algonquian origin and translates as "eaters of raw meat." To the Inuit, the word Eskimo is offensive, since it congers hurtful stereotypes and romanticizes a mythic rather than accurate lifestyle.

The Inuit live in the treeless, frozen landscape of the Arctic Climate Zone extending from Greenland and the northernmost parts of Eurasia and North American continents, which includes regions of Canada, Alaska, Siberia, and Scandinavia. There is a large amount of variation in climate across the Arctic, but all of the regions experience extremes in solar radiation in summer and in

DOI: 10.4324/9781003041627-5

Figure 3.1 Illustration of Inuit from a voyage by Martin Frobisher in the 16th century. From: Beschreibung der Schiffart des Haubtmans Martini Forbißher, Nuremberg 1580. Unknown author. Public domain. Wikimedia.

winter. Some parts of the Arctic are covered by ice throughout the year, and nearly all of the Arctic has long periods with ice and snow on the surface.

Current average temperatures in January range from about −29 to +32 degrees Fahrenheit, and winter temperatures can drop below −58 degrees F over many parts of the Arctic. The coldest recorded temperature in North America was in Snag, Yukon on February 3, 1947, plummeting to −81.4 degrees F. Current average July temperatures range from about 14 to 50 degrees F, with some areas periodically exceeding 86 degrees F in the summer.

The Arctic Climate Zone consists of ocean that is largely surrounded by land, and thus climate in many parts of the Arctic is influenced by ocean water. The zone is subject to long, severe winters with few hours of daylight and short summers with almost endless daylight. The ground has a thick subsurface layer of soil—permafrost—that remains frozen throughout the year. It can be an inch or over miles deep. There is less precipitation in the Arctic than further south, but snow falls regularly stirred up by frigid winds resulting in strong blizzards and vast drifts. Some northern waters, even parts of seas and oceans, freeze over during the winter, with ice floes during the summer thaw.

Yet the Arctic Climate Zone is more varied in its terrain and seasonal diversity than one would imagine. Anja Nicole Stuckenberger in *Thin Ice: Inuit Traditions within a Changing Environment* aptly stated, "The landscapes of the Arctic zone vary remarkably during the year, alternately snow-covered and grass laden, blanketed in permafrost and filled with wildflowers, thick with frozen sea ice and rolling with slate-blue waves" (Stuckenberger, 2007, 29). There are tundra (Sami for "barren land") regions throughout the zone with freezing temperatures, a lack of trees, low-growing vegetation, and numerous rock outcrops. Despite the severity of the environment, there is an abundance of plant and animal life on the tundra. Among the variety of plants are lichens, mosses, grasses, and low shrubs. Numerous mammals, birds, fish, reptiles, and amphibians live there in the summer, but many migrate in the fall. More than 3,000 different insect types also live there, as well as various bacteria and fungi.

The Inuit religion, spirituality, and mythology were shaped to a large extent by the environment in which they lived. The central precept of their religion was that nature essentially was malevolent. They must placate nature's forces and observe strict taboos. Prior to contact with Europeans, shamans (*angakuit*) were the Inuit religious leaders who underwent exhaustive training. They were intermediaries between the Inuit and a variety of spiritual forces that influenced them. In Inuit mythology, there is a belief in other worlds beneath the seas, inside the ground, and in the sky, where gifted shamans could travel via trances and in dreams.

The Inuit lived with an understanding that much of what they experienced was uncontrollable, which made it all the more important that they followed a set of rituals in daily life. They believed in an afterlife, and that all things—including animals—had a soul or spirit. They were reverential of life and the need to treat the animals they hunted with respect. It was common for hunters to open the skulls of freshly killed animals to release their spirits. As one observer noted, the Inuit did not worship a divine figure, but they feared much.

Above all, Inuit understanding of climate and weather reflected their largest belief systems. The idea of *sila* is the closest to a modern notion of climate in Inuit culture, connoting "universe," "sky," and "weather." According to Stuckenberger, "*Sila* expresses itself in the changing seasons, which in turn shape the Inuit long-term expectations of, for example, weather conditions, snow and ice quality, and periods of open and frozen sea ... In the pre-Christian beliefs of the Inuit, *sila* was associated with a spirit master named Narssuk, a giant infant who was unpredictable in behavior and temperament, just like the Arctic weather" (Stuckenberger, 2007, 33).

The Inuit and their predecessors have occupied the Arctic for about 4,000 years or more, and today most of the Inuit continue to live there, including in coastal settlements throughout the islands. We were likely to find the earliest ancestors of modern Inuit living in small communities along the coastline of the Bering Land Bridge. As the population grew, settlements spread north along the coast and possibly inland via the large river valleys and extended

Figure 3.2 Inuit people (from a book published in 1906). Unknown author. Public domain. Wikimedia.

farther north into Alaska and as far as Greenland. The weather became less hospitable as they moved north with thick ice covering the sea in winter.

This led the settlers to shift their way of life to this seemingly unlivable environment. The Inuit and their forebears did this by gaining a deep knowledge of the sea ice and developing the necessary skills and technology to house themselves, travel when needed, and successfully hunt and fish for the native animals by observing their behavior.

There is an old maxim in popular mythology that the Inuit (usually referred to as Eskimos in the story) have something like 52 words for ice and snow. Linguists and anthropologist, however, challenged this assumption. Lucien Schneider, a linguist and missionary, claimed that in Inuktitut (one of the principal Inuit languages spoken in Canada) there were a dozen basic words (not derived from another word) referring to snow and about ten words referring to ice. Yet the Inuit also have several other ways to describe ice and snow not limited to a single word. In Inuktitut, new words can easily be created from base terms (an agglutinative language). For example, *qanik* refers to falling snow, but *qanittaq* (added snow) refers to fresh falling snow. *Siku* means ice, in general, while *sikuaq* is the first layer of thin ice forming on puddles in the fall. While linguists continue to debate the way the Inuit express

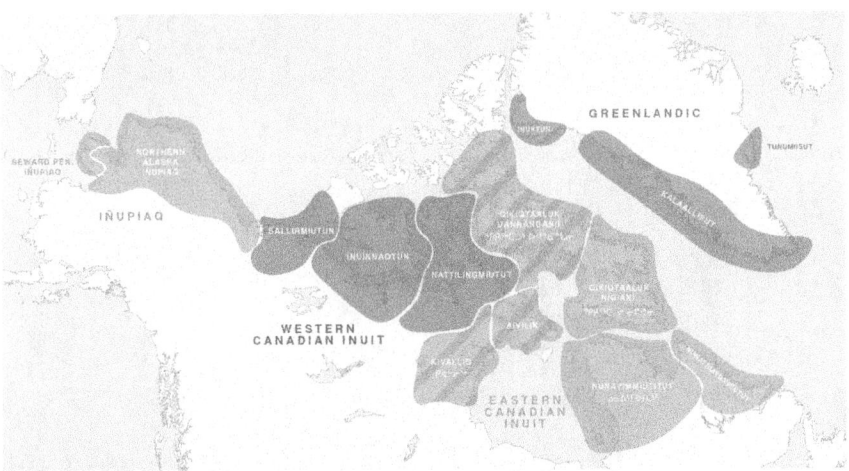

Figure 3.3 Inuit languages and dialects.

Sources: http://www.languagesgulper.com/eng/Eskimo_files/droppedImage.jpg.

http://www.ya-native.com/Culture_Arctic/image/articmap.png.

https://www.uaf.edu/anlc/images/ipla-map-20130712_sm.jpg Creative Commons Attribution-Share Alike 4.0. Wikimedia.

snow and ice, the bottom line is that these people have demonstrated a superior—and often subtle—way to express their environment in words. In addition, the Inuit approach to sea ice is based on experience, not theory.

Sea ice in particular, and ice and snow in general, are integral to the Arctic lifestyle of the Inuit. Their many terms for these natural phenomena speak to a world that they encountered and dealt with daily. The places where sea ice meets open water are vital to Inuit livelihood. In essence, ice is an extension of land, allowing the Inuit to travel and gain access to various communities and to harvesting areas—exploiting resources associated with it. According to zoologist Rick Riewe, "In the Arctic, sea ice develops during the autumn and winter months and exists as a transitory environment until mid-summer. In the High Arctic, this winter phenomenon exists year-round, and multi-year ice commonly occurs. It is in this winter marine environment that the Inuit evolved. Over most of the Inuit range, their culture and language have been intimately and inextricably linked to the sea ice and the marine food webs" (Riewe, 1991, 8).

Sea ice (frozen water with pockets of brine, which includes habitat for bacteria and algae), for example, varies in type and form depending on climatic and other environmental conditions. Despite being solid, it is less dense than seawater and consequently floats. Much of the ice away from land (drift ice) is constantly in motion, driven by winds and ocean currents. There are several types of sea ice: newly formed ice, *nilas ice* (less than about 4 inches thick), *young ice* (about 4 to 12 inches thick), *first-year ice* (more than 12 inches thick); and *old ice* (ice that survived a whole season or more).

The forms of sea ice also vary. Small *pancake ice* is almost 10 feet, whereas *ice floes* can be more than 65 feet. *Pack ice* usually is composed of floes greater than 10 feet and moves with the currents and winds. *Fast ice* (*land-fast ice* or *tuvaq*) is attached to the shoreline, shoals, or grounded icebergs and is generally stable. The seaward side is a habitat for breeding seals, and a hunting area for humans, polar bears, and Arctic foxes. There are areas of open water within sea ice referred to as *leads* or *polynyas*, kept open by currents, tides, and wind.

From the Inuit's point of view, according to anthropologist Claudio Aporta, the dynamics of sea ice "are understood to the deepest degree, to the point that, to experienced hunters, walking on thin ice, and dealing with such events as moving ice are not extraordinary feats, but ordinary ways in which they interact with this environment. Such knowledge has been developed through generations of careful and systematic observation of sea ice behavior." Furthermore, Aporta states, "To Inuit, sea ice possesses a social life ... and is seen and perceived as home." This is quite different from the view of European explorers, who viewed the stretch of water with sea ice to be "a transitory place or a transportation medium" (Aporta, 2011, 7–8).

In very practical ways, knowledge of sea ice (and respect for it) expands Inuit territory and provides access to many resources. The Inuit's understanding of the ice comes not only from deep knowledge but also from sight and sound. The shifting of sea ice can be treacherous. In precontact times, for example, a hunter in the Point Hope area (along the Chukchee Sea) who was less than twenty-five years old was expected to hunt under the guidance of an older, more experienced man.

The Inuit know well the ice areas which move away and return to expose places and resources with which they are familiar. They understand sea-ice topography in the same way others understand rivers, lakes, rocks, hills, and mountains. They recognize relationships among the phases of the moon, wind direction, tidal currents, geography, animal behavior, and the actions of ice. In this way, the knowledge the Inuit accrued helped them anticipate change, predict where animals will be located, and minimize risks. For non-Inuit, especially European explorers, sea ice was considered an impediment to travel, and never part of a habitable environment.

Because the Arctic region was home to the Inuit, their shelter and livelihood adapted to the harsh conditions present in daily life. Housing varied by season, location, and purpose. Archeological information is limited, but some patterns emerged. Sivullirmiut (precursors to the Inuit) lived in small groups. Their camps were located close to where they easily could find animals to hunt, and depending on the season, where they could gather clams and mussels, seaweed, bird eggs, and berries. In the summer they likely lived in skin tents; in the winter they probably employed partly underground houses with walls of stone or sod. These groups may have built snow houses, but no artifacts remain to tell for sure. On the contrary, Thule (another precursor to the Inuit) winter villages were quite large, maybe six to thirty houses made of stone slabs and cut sod supported by whalebone framing. The nature of whale hunting likely dictated the size and type of the built community.

Some Inuit lived in log cabins insulated with turf or sometimes in a temporary dwelling made of ice, generally known as an *igloo*. Often, the Inuit left larger winter villages for ice camps in the spring to hunt and fish. In the early 1910s, archeologists found abandoned Inuit villages on the sea ice along the Dolphin and Union Strait in Canada. Each of the villages was believed to consist of about 50 *igluit* or *iglooit* (plurals of igloo). There was, however, no evidence of other types of building materials, but by the process of elimination, the assumption was that snow houses had been present. The location was determined by proximity to hunted seals (probably no more than about 5 miles from camp).

The igloo—or snowhouse—is probably the most recognizable artifact associated with the Inuit. In many cases, they largely were seasonal dwellings; when temperatures rose above freezing tents made of animal skins and bones often replaced them. In some cases, they were used only when traveling. The snow shelter was constructed out of blocks of ice, primarily in the form of a dome, built spirally from within. There are different sizes of igluit, serving different purposes. The smallest igloo normally was a temporary shelter for hunters. There also was a larger size igloo for family use that could be semi-permanent. The largest igloo was built in groups of two—one for living, another for special occasions. It might house 20 people in as many as five rooms attached by tunnels. Other Inuit used snow to insulate a house made of whalebone and hides rather than relying on ice blocks.

The early Inuit made their livelihood by hunting and fishing and applying the knowledge, techniques, and technology acquired in their icy surroundings. Food gathering also was essential for bird eggs, shellfish, and berries. Since fresh food was not always available, the Inuit preserved, dried, and stored some food items. Inuit society normally had a division of labor with men generally engaged in hunting and fishing, and women involved in caring for children, cleaning and sewing, processing food, and cooking. However, the gender roles, while strong, were not necessarily absolute. There were situations where women hunted out of necessity or personal choice, or men who needed to know how to cook and sew when away from camp. A small number of family units normally made up a band of Inuit. In some groups, there was an informal leader (*isumataaq*), who could be a young man who was regarded as an effective hunter and who displayed leadership skills.

There could be between 12 and 50 people in a hunting party. Hunting seasons revolved around the time of year, and usually focused on one animal at a time, such as seals, walrus, whales, or caribou. They also hunted polar bears and several smaller mammals. There was some variation in practices and the types of quarry hunted in different parts of the Inuit world. For example, the Arvirtuurmiut of Boothia Peninsula in Canada were regarded as "baleen whale-eating people." In Labrador, while coastal waters were not iced over, the men hunted walrus, seals, and beluga from kayaks. In late autumn they hunted bowhead whale (a baleen whale of the Arctic seas). In the winter, they tracked seals near the edge of the ice. On the moving ice in various locations, the Inuit concentrated on walrus.

The tools utilized for hunting, fishing, and other purposes often were ingenious—some borrowed, some invented—but all adapted to the environment in which the Inuit lived. The Sivullirmiut made small, delicate tools out of antler, ivory, bone, and driftwood. The Thule hunted a variety of mammals, including large whales in the Arctic Archipelago.

Some of the Inuit's most essential tools included harpoons, fishing spears, lances, bows and arrows, and even at some point snow goggles. The Inuit employed many of these items and often adapted them to their own use. They also utilized hides and bones for non-weapon purposes, such as for plates, water pails, parkas, and other protective clothing, and grass and whale baleen for baskets and other containers. The tools and weapons utilized by the Inuit grew out of their environment and evolved from older and newly devised technologies.

The available means of transportation were particularly functional and water- and ice-dependent. Snowshoes are associated with Arctic people, but these were not used universally. For more than 2,000 years the Inuit (except for the most northerly) used the sealskin kayak—*umiaq* or *umiak*—for traveling and hunting. The umiaq was normally a single-person, closed-deck boat of about 13–23 feet in length (which could hold three people if necessary). Wooden paddles propelled it, which allowed the umiaq to move quickly in the water.

On land or long stretches of ice, the dogsled was a preferred method of travel in the winter. Teams of from two to twelve or more dogs were tied to the sled in pairs (sometimes in single file between two towlines depending on conditions). Sleds varied in type. In the Arctic, the Inuit developed the *qamutik*, a heavy sled designed to carry loads in rough terrain. A flat-bottomed toboggan might be used farther south. Dogs were used as hunting animals to locate seal breathing holes in the ice, and for hunting muskoxen and bears. Dogs also might serve as pack animals in the summer.

The ability of the Inuit to adapt their culture, their living environment, their hunting fishing prowess, and their transportation to the harsh weather and to the sea ice was impressive. After contact with Europeans, however, their way of life persisted, but much changed. The Vikings under Eric the Red encountered the Inuit in Greenland in 984. About six hundred years later, the British explorer Martin Frobisher found Central Inuit in northern Canada (including Baffin Island), and other missionaries, explorers, and whalers made contact with the Inuit in southern Labrador. In 1741, the Russian explorer, Vitus Bering, met the Inuit of Alaska. Unlike Inuit, Aleut, and other groups living below the Arctic Circle, northern Inuit were not immediately affected by contact with Europeans. Over time all that changed—sometimes violently, sometimes through trade, and often through religion.

Christianity competed, modified, or ultimately changed Inuit belief systems. Over many years, Inuit ways of life were threatened by the political, economic, and social changes that came with colonization and non-Inuit settlement in *Inuit Nunangat* in the form of overcrowding, food scarcity, chronic health issues, and even high rates of youth suicide.

The major transformation of Inuit lifestyle in recent years was not foreign contact or even Western religion, but the intensification of climate change. The Arctic region is much warmer than it used to be and it continues to get warmer. In the last several decades, it has heated more than any other region in the world. Human-induced climate change through the use of fossil fuels and other elements has caused and will continue to cause changes to sea ice, snow cover, and the extent of permafrost in the Arctic.

Sea ice is an important part of the climate system of the Arctic. Although it appeared to be a major obstacle to human life, it is responsible for insulating the ocean from the deathly cold of winter. In the summer it increases the surface reflectivity over that of the open ocean. An increase in ice, therefore, decreases the total amount of solar radiation absorbed by the water, thus reducing surface temperature and allowing more ice growth.

It is the reverse for a decrease in ice cover. Also, the transport of drift ice outside the Arctic provides a source of fresh water to the North Atlantic. In 2003, research scientists Gerhard S. Dieckmann and Hartmut H. Hellmer stated, "Today we know that the annual cycle of sea ice formation and degradation not only plays a pivotal role in governing the world climate but also influences processes in the oceans down to the abyss" (Thomas and Dieckmann, 2003, 1).

Across the North, there are additional problems associated with coastal erosion, thawing permafrost, and serious runoff. A preview of an Arctic Climate Impact Assessment, presented to the World Conference of Science Journalists in Montreal recently reported that many marine species such as walrus, polar bears, and seals could go extinct by 2070–2090.

Adaptation for Inuit is a difficult path affecting hunting and fishing practices and access to wildlife, food storage, transmission of diseases, accidents due to ice melts, impact on existing infrastructures of all types, transportation access, and even increased tourism. It is difficult, in addition, for elders to hand down to the next generation beyond their knowledge of weather patterns, ice floes, and everything else they have learned about the environment as it changes rapidly. Duane Smith, President of the Inuit Circumpolar Conference and Vice-President of Inuit Tapiriit Kanatami stated, "In the last 40–50 years, Inuit have adjusted to social, economic, and cultural changes. But even as we adapt to globalization, we realize that climate moderation is likely to be the key driver of socio-economic and cultural changes in years ahead" (Smith, 2021). Sheila Watt-Cloutier, an Inuk, is a respected political figure in the Arctic. In 2018, she said, "Because temperatures in the Arctic are rising faster than anywhere else in the world, we must look to the experiences of Inui as a harbinger of what is to come, and seek their guidance on how to live more sustainably" (Watt-Cloutier, 2018).

The history of the Inuit has been defined by its relationship with the hostile climate and weather of the Arctic to which it adapted successfully in many ways for many years. Can they adapt to an uncertain future challenged by climate change and pass that experience on to the rest of us?

For generations in that hostile northern environment, the predominant element was frozen water. Other cultures in other hostile places tried to survive and adapt by confronting water in different contexts and of different sorts. Water as a liquid—fresh or salty—poses its own set of challenges, but contending with frozen water is particularly vexing. Yet the Inuit and other northern peoples found methods to utilize water in that form, not unlike the Aztecs who adapted to their terrain. The necessity for water—and frozen water at that—inspired the Inuit to develop a lifestyle that undergirded their very culture. The future of the frozen water and the environment in which it dominates is being transformed in ways that the Inuit and the rest of us probably never imagined. What will be the new normal that must be confronted?

Notes and Further Reading

Aporta, Claudio (2011) "Shifting Perspectives on Shifting Ice: Documenting and Representing Inuit Use of the Sea Ice," *Canadian Geographer* 55: 6–19.

The Canadian Encyclopedia (2013) October 2013, online, https://thecanadianencyclopedia.ca/en.

Dowsley, Martha, et al (2010) "Should We Turn the Tent? Inuit Women and Climate Change," *Inuit Studies* 34: 151–165.

Riewe, Rick (1991) "Inuit Use of the Sea Ice," *Arctic and Alpine Research* 23: 3–10.

Smith, Duane (2021) "Climate Change In the Arctic: An Inuit Reality," *UN Chronicle*, accessed October31, online, https://www.un.org/en/chronicle/article/climate-change-arctic-inuit-reality.

Stuckenberger, Anja Nicole (2007) *Thin Ice: Inuit Traditions within a Changing Environment* (Hanover, NH: Hood Museum of Art, Dartmouth College).

Thomas, David N., and Gerhard S. Dieckmann, eds. (2003) *Sea Ice: An Introduction to Its Physics, Chemistry, Biology and Geology* (Oxford, UK: Blackwell).

Vanderzwaag, David, and Donat Pharand (1984) "Inuit and the Ice: Implications for Canadian Arctic Water," *Canadian Yearbook of International Law* 21: 53–84.

Waldman, Carl (2006) *Encyclopedia of Native American Tribes* (New York: Facts on File, 3rd ed.).

Watt-Cloutier, Sheila (2018) "It's Time to Listen to the Inuit on Climate Change," *Canadian Geographic*, November 15, online, https://www.canadiangeographic.ca/article/its-time-listen-inuit-climate-change.

Part II

Colonialization and Early-Industrial Growth

The essays in this section confront three transformative technologies. The first deals with the development of acequias and Spanish water law beginning in the sixteenth century when Spanish and Mexican explorers and settlers entered the American Southwest. The next essay discusses the origins of commercial fishing in Newfoundland beginning in the "Age of Explorers" in the fifteenth century when several European countries sought to exploit the vast fishing wealth on the Atlantic coast of modern-day Canada. The final essay treats the impact of the waterwheel and then the steam engine primarily in the eastern United States in the period of early industrialization.

While the topics are disparate, they share a common feature: Each explores the connection between water and forms of technologies which produced economic, social, and environmental change.

The Spanish were the first Europeans to establish a system of irrigation—*acequias*—in what became the American Southwest. It was, however, based on Native American and European traditions and technologies. Not only did the system revolutionize agriculture in this arid region, but the acequias were governed at the community level. Access and use of water were determined by collective decisions, not by the leaders alone or by the countries they may have represented. The resulting water law imbedded in it this communal right to water. Over time, competing concepts of water law undermined this approach, however. Nevertheless, the acequias infrastructure dominated the southwestern environment, transforming lands, attracting settlers, and spawning successful agricultural production.

Commercial fishing in Newfoundland also attracted Europeans to North America, not in search of cities of gold, but seeking maritime resources not as plentiful in their homeland. Cod and other fish and shellfish along the Atlantic seaboard offered high-quality protein that traveled well after processing. A variety of technologies were employed to access, ship, and utilize this resource. Fishing techniques evolved over time from hand-fishing and netting to more efficient and larger scale approaches to "mining" fish. Methods for preparing and preserving fish for travel also evolved out of local experiences and ideas originating in Europe and elsewhere. Later, refrigeration made transporting

DOI: 10.4324/9781003041627-6

fish much easier. Even the types of boats were adapted to the hostile marine environment that the fishers experienced.

The commodification of fish also inspired conflicts within the industry and among the various countries vying for the biggest and best catch. The demand for fish and efforts to dominate the market led, in some cases, to overfishing and required regulatory measures to maintain supplies. A number of externalities—pollution, climate change, economic challenges—all could impact what seemed to be a never-ending abundance of a resource highly in demand.

The invention and utilization of the waterwheel and the steam engine changed agriculture and industry alike. There was nothing static about these technologies, which were found changing all of the time to meet a variety of needs and wants. They shared the objective of turning water into mechanical energy with different levels of success.

Waterwheels became more sophisticated over time but never could escape the fact that they were bound to places where water flowed. American communities in the East, South, and parts of the Midwest were more amenable to waterwheels than the arid West. And while building waterwheels could transform watercourses in significant ways, the mobility of the steam engine—its major advantage—resulted in creating billowing smoke almost anywhere. Waterwheels depended on running water and clever technology to create mechanical power, but steam engines needed fresh water, wood, and/or coal for it to operate. Steam engines essentially introduced fossil fuels into industrial society.

In this period, water was a key element of commercial and industrial change. In acequias, water was directed to make land fertile. Access to the marine environment provided vast protein for the table. Waterwheels and steam engines transformed one source of energy into another that could be used for many kinds of work and production.

4 Acequias and Spanish Water Law

The Spanish were the first Europeans to establish a system of irrigation—*acequias*—in what is now the United States. Acequias are community irrigation systems found in the villages and pueblos of New Mexico and the Southwest (also in northern Mexico and the Andes). What makes the acequias historically important—especially with respect to Spanish water law—was how water sources and supplies were to be treated as commons. Speaking about New Mexico, but applicable to the American West, water expert John R. Brown asserted, "New Mexico's diverse Native American and Hispano *acequia* traditions both inform and complicate the process of crafting institutions for governing the water resources of the state" (Brown, 2004, 1).

The acequia tradition of governing them at the community level and making collective decisions about access and use of water were imbedded in some of the practices and laws in the West, but not in others. Spanish water law would be influential, but also highlighted contests and conflicts over water rights and water governance.

Spanish/Mexican settlers came from Europe and central Mexico (Mexico City, Guadalajara, and Zacatecas), entering the upper *Rio del Norte* (Rio Grande or Rio Bravo) of the northern reaches of *Nueva Espana* in the late sixteenth century. In this northern sector of the viceroyalty, they penetrated places such as Nuevo Leon and Coahuila, Mexico; Sonora, Arizona; Baja and Alta, California; various locations in Texas; and Chihuahua, New Mexico. In New Mexico, they encountered an area occupied by Pueblo Indians that was rich in natural and mineral resources, but short on water as elsewhere in the desert regions of the Southwest.

Aridity dominated the Southwest. The northern boundary of Nueva Espana in the sixteenth century ranged from extremely arid to semi-arid, except for the eastern part of Texas. Yet, it was more varied than the colonials' homeland. According to historian Michael C. Meyer, "It had a wider range of altitudes, soils, animal life, drought-resistant vegetation, and even more capricious cycles of annual rainfall. The mountains were more rugged and towering, and the barrancas or canyons more impenetrable. Erosion and sedimentation bequeathed a physiography at once both harsh and captivating–frightening, yet alluring…" (Meyer, 1984, 22).

DOI: 10.4324/9781003041627-7

Although primarily desert, there was a rainy season from July to September with few areas exceeding twelve or thirteen inches per year; some receiving rainfall less than seven or eight inches. The mountains in the region captured most of the moisture originating in the winds from the Pacific Ocean and the Gulf of Mexico, and from winter snow cover. Water flowed in modest amounts through the area's small rivers, including the Rio Grande, the Colorado, the Fuerte, the Yaqui, and the Gila. None of them were particularly useful for transportation nor commerce, and most were not perennial, that is, running only part of the year. Despite their limits, the rivers drew people and wildlife.

In 1539, Marcos de Niza led an exploration north from Mexico to the Zuni pueblos. The next year Francisco Vasquez de Coronado traveled as far east as present-day Kansas. In 1582, the Spaniards explored many pueblos including the Zuni, Hopi, Acoma, Tiwa, and Keres. They returned via the Pecos River in New Mexico to avenge the killing of Franciscan friars, who had been left there by the previous exploration party.

Don Juan de Onate established the first Spanish colony in New Mexico in July 1598 (San Juan Pueblo) with about 400 settlers, several Franciscan missionaries, and 7,000 head of cattle, sheep, and horses, and almost immediately began working on an irrigation canal. Soon he constructed the town of San Gabriel on a partially abandoned Tewa pueblo, where Acequia de Chamita was constructed in about 1600. This was possibly the oldest, continually functioning irrigation ditch of Iberian origin in New Mexico. San Gabriel remained the capital city of the province until 1609–1610 when it was moved to Santa Fe.

The settlers were not easy on the local tribes. They made them pay taxes on cotton crops, cloth, and work, and Onate also treated them brutally for which he was tried by a Spanish court. This was only the first time that the Spanish colonizers treated the indigenous peoples cruelly. Although the Franciscans meant to strip the local tribes of their native religion and cultural ways by converting them to Christianity, they also provided a measure of protection from the violence.

There were several incidences that led to unrest, including the Great Pueblo Revolt in 1680, spurred on by the establishment of the *encomienda system* meant to define and control the role of the locals. The Pueblo were able to push the colonists out of New Mexico, only to have them return by capturing Santa Fe in 1692. The relationship improved somewhat at this time. Disease, domestication of animals, and exposure to new weapons, however, changed the life of southwestern indigenous peoples by the end of the seventeenth century.

The Spanish conquistadors and the missionaries chose their first settlements in *El Reino del Nueva Mexico* close to established pueblos because of the potential water sources, the people available as laborers, and the number of new souls to convert to Christianity. Early exploration maps of the area showed the locations of rivers, creeks, lakes, and even small water features such as ponds and dry arroyos.

The European and indigenous cultures came together in the building of a vast system of irrigation in the Southwest including New Mexico, Arizona, California, and Texas. Despite some cultural borrowing, the intent of the colonists was not to adapt to local ways but to further their own interests. As Meyer stated, "The first Spaniard in the Southwest, like those elsewhere in the vast dominions of the Spanish empire in America, sought to establish a homogenous society, rooted in Spanish tradition and secured by the unquestioning allegiance of both the Crown and the Roman Catholic Church" (Meyer, 1984, 3).

It just so happens that the Spanish and indigenous peoples shared some common traditions when it came to developing practical water systems. And, at first, diverging interests and competition in securing and sustaining water rights did not seem to be at stake. Generally, when a new Spanish town, presidio, or mission was founded, existing native communities were guaranteed a share of the water supply.

Figure 4.1 Acequia Madre de Valero (Main Irrigation Ditch of Valero Mission). One in a network of ditches begun by the Spanish and indigenous people at the founding of San Antonio in 1718. This section was restored in 1968. Darryl Pearson. Creative Commons Attribution-Share Alike 4.0. Wikimedia.

Because of the early settlement there (and effective agricultural development), New Mexico (*La Provincial del Nuevo Mexico*) was the initial focus. It sits at the intersection of four geologic areas: the Colorado Plateau, the Rocky Mountains, the Great Plains, and the Basin and Range Province. In the deep past, it was wet and lush, but it became a desert at the time of the first human settlements in 10,000 BCE. Water was scarce, with only six modest rivers.

Most acequias are open ditches with dirt banks, but some can be gravity chutes. The Spanish word *acequia* is related to the Catalan word *sequia* which comes from the Arabic *as-saqiya* (root *saqa*), meaning "to irrigate." Acequias evolved over 10,000 years in the deserts of the Middle East and were introduced in southern Spain by Moors during their 700 to 800 year occupation. After the Moors were driven out of Spain in 1492, their hydraulic systems nevertheless remain in places such as Valencia, Murcia, and Andalucia. Before the Moorish occupation of southern Spain, the Romans introduced their agrosystems, knowledge of water distribution, various technologies, and laws related to water in Spain. The Romans first came to the Iberian Peninsula in 206 BCE, which gradually led to conquest in the following decades.

The colonial acequias in the New World were an amalgam of Spanish/Moorish and indigenous water systems and practices. Settlers from the interior of Mexico and Spain had observed the practices of the locals as they established their own bases throughout the Southwest. Ultimately, the new settlements produced what some have called "a hybrid legacy." By 1800 there were 164 acequias in the upper part of the Rio Grande River basin.

At the time of Spanish settlement in the Southwest, Pueblos of New Mexico were leading all tribes there in harnessing water for irrigation (Navajo were active as well). Sites on the upper Rio Grande Valley since about 1400 (with terraces and reservoirs) were producing maize, squash, beans, melons, cotton, and chili. The technology was not elaborate but mirrored what the Europeans had experienced at home along the Mediterranean.

As we have seen earlier, the Pueblo were not the first indigenous people in the arid Southwest to develop and utilize irrigation. By the time the Spaniards began their conquests in Mesoamerica, the Hohokam were gone and their canals filled with sand. Also gone were the Anasazis ("ancient ones") who occupied the Four Corners area (southwestern corner of Colorado, southeastern corner of Utah, northeastern corner of Arizona, northwestern corner of New Mexico) about the same time as the Hohokam. Prior to 1500 BCE, the Anasazi were hunter-gatherers, but as their population grew, they needed more permanent and larger quantities of food. The Anasazi planted maize, squash, and beans on contoured terraces, bordered gardens, and on the canyon floor in the high desert area where they lived, insuring better land and water conservation. Water came from precipitation and runoff from the tops of mesas.

Near Mesa Verde (southwest Colorado), the Anasazi built check dams (small, sometimes temporary dams) and an irrigation ditch of about four miles in length. They captured flows during rainstorms via check dams to use for flooding agricultural fields. At Chaco Canyon (northwest New Mexico), they constructed check and diversion dams with canals and head gates.

Because of drought and other factors, the Anasazi began to abandon the Four Corners area, and by 1400 BE had relocated principally along the Rio Grande Valley. The lower altitude and more reliable sources of water made possible new settlements for the Pueblo Indians of New Mexico (made up of

Tewa, Tiwa, and Keresan tribes), who were the descendants of the Anasazi. The Pueblo continued floodwater farming practiced by their ancestors, but also utilized dry farming and communal systems of irrigation. They diverted water flows along arroyos, creeks, and tributaries and then channeled water by way of wide and shallow canals to cultivate their fields. As professor of regional studies Jose A. Rivera and Spanish counselor officer Luis Pablo Martinez stated, "When the conquistadores encountered these ditches, they marveled at the complexity of some of these systems while noting their resemblance to Spanish irrigation canals" (Rivera and Martinez, 2009, 318).

Tensions between the colonial Hispanos and Pueblo took time to fade. However, the two peoples shared not only a reverence for water but a belief in community water rights and a commitment to community-based irrigation practices. The communal system of irrigating was a logical response to water scarcity in the region and a mechanism of survival for the agricultural communities. The acequias in New Mexico were the oldest water management institutions of Spanish-Indian origins in the United States.

Going by several names—community acequia, public acequia, or community ditch—an irrigation organization was composed of the owners of lands bordering on it or irrigated by the ditch. This group constructed and maintained the acequias. Property owners were required to participate in maintenance and upkeep of the acequias in direct proportion to the benefits derived from the irrigation system. The owner/irrigators functioned ideally as *comunidades de regantes* or "water democracies" because they were meant to be autonomous and operated mostly outside of government with respect to internal affairs. They elected their officers, established and enforced rules, and settled water disputes. In some cases, a different management system formed in communities that had little or no legal status and had no town government. The majority of settlers lived in small rural communities within which the community ditches were developed voluntarily by water users. Distribution was regulated by an elected boss or *mayordomo*. In some cases, acequias were the property of individuals or jointly shared by a corporation of rural users.

As we have seen, early infrastructures using water were a principal tool of landscape modification to serve the needs of humans. The building process was not easy, and problems varied from place to place. The quality of the water (there were no water standards at the time), the volume of water, the surrounding landscape, and climatic factors all influenced the success of a project. In a larger sense, many urban landscapes in the Southwest were strongly influenced by the developing water systems. Cities like Sant Fe, Albuquerque, San Antonio, El Paso, Tucson, and Los Angeles were settled under colonial law. "Although the modern utility and use of the acequia irrigation system varies from each city in the region," stated community and regional planning professor Moises Gonzales, "it is certain that the urban form expressed in these cities was generated by the acequia cultural landscape" (Gonzales, 2014, 894). This system became the foundation for organizing several urban areas in large and small cities alike.

Building acequia irrigation after the arrival of the Spanish/Mexican settlers relied heavily on local labor (mostly under duress), and in several situations could be greater in scale (and impounding capability) and more complex than their predecessors. The building of the acequias required a great amount of labor-intensive work as had been true for the ancestors of the locals. Ditches were dug with wooden spades and contoured with knives. Rawhides drawn by oxen or suspended on poles were used to move the dirt.

In many locations, acequias were built at the same time or even before a local mission, church, or presidio. The first step was to locate a bend in the river or another appropriate feature to build a diversion structure (out of timber, brush, and rocks) to capture water and deliver it to the ditches. Gravity flow was used to move the water. On large streams, workers built wing dams (extending part way into the water) along a bank to channel water into the acequias during irrigation season (when flows were highest). On smaller streams, they constructed dams (*presas*) to create small reservoirs. Afterwards they excavated the main canal, or *acequia madre* (mother ditch), off either or both banks. In most cases, the *acequia madre* was located perpendicular to the stream source towards the upper end of the community and then moved water downstream parallel to the river. After flowing along the slope of the terrain for several miles, water was returned to the original stream through a drainage channel (*desague*).

Aside from the practical nature of the design, the acequia defined the borders of the community. It was well understood that this common property ditch, which offered shared ownership and joint responsibility, was impossible for any single cultivator or irrigator to build or maintain. The irrigation system also formed a greenbelt—or a kind of oasis or microclimate—along its course, providing and sustaining habitats for plant diversity and for wildlife. By returning surplus water by way of desague, the acequia recharged local aquifers. By 1700, approximately 60 acequias were operating in New Mexico alone, with another 100 during the next century. In the 1800s at least 300 more acequias were built in New Mexico.

Under the best of circumstances, Pueblo and Hispano acequia communities maintained local control over water and developed customs and practices concerning allocation (challenged in the nineteenth century). What the Spanish brought with them that indigenous people did not possess —aside from new technology, new economic interests, and new crops—was a body of formal, written water law that created a unique legal structure related to water in New Mexico and elsewhere in the Spanish/Mexican-controlled Southwest.

Developed through ancient Roman and Islamic law, Spanish water law was unchallenged until at least 1848 with the end of the American war with Mexico, and among other things, was a way of maintaining dominance over native practices even while absorbing many of them. As environmental and legal historian Kathleen A. Brosnan stated, "West of the Mississippi, for nearly three centuries prior to 1848, Spanish water law dominated much of the region" (Isenberg, 2014, 526).

During Spanish rule, the crown retained ownership of their colonies and their water, mineral, and land rights. But the legal system provided great latitude in responding to local needs. A basic principle was that surface water be shared as a common resource. Irrigation was seen, not as a way to create ownership of water or to encourage individual enterprise, but to establish a permanent community. When conflicts arose—and they did—the courts attempted to render "the greatest good for the greatest number."

Spanish legal systems continued without much modification in New Mexico, Texas, California, and elsewhere in the West and Southwest even after the formation of the Republic of Mexico in 1821. (Mexico ruled Texas until 1836 and California and New Mexico until 1848.) However, a series of water laws during the Mexican period of government added specific penalties for various violations. For example, someone who bathed in or polluted water in other ways was subject to a fine.

After the Spanish conquest of Mexico and Peru, the general principles of Spanish jurisprudence were incorporated in a legal code written for the Spanish empire called the *Recopilacion de las leyes de los reynos de las Indias* (Laws of the Indies), which established the foundation for Spanish water law in the Americas. Through the Laws of the Indies, ownership of all land, water, and minerals were vested in the crown. With respect to the community water systems of the indigenous peoples, the Laws of the Indies generally recognized that the local laws and customs should be retained and respected.

Surface water and groundwater were treated differently under the law. Surface water was *propiedad imperfecta*, meaning that it was measured against the rights of others. Spanish property law did not recognize riparian rights (relating to or situated on the banks of a river or stream) to running water. If property fronted on a river or stream, the owner only could use water for domestic purposes. However, the crown could confer surface water rights for agricultural and industrial purposes through a specific grant. In the northern frontier of New Spain, a *repartimiento de aguas* was used to convey water rights, but it was more common in most of the colonial region for surface water rights to be granted automatically if the land was determined to be irrigable. In the case of groundwater (*propiedad perfecta*), possession of spring water or rainwater most typically fell to the landowner.

As American settlers began to inundate the Southwestern territory, it was unclear if the existing water laws would survive. The first Anglo-Americans to embrace the possibility of large-scale centralized irrigation technology were Mormon refugees (or Latter-day Saints) who emigrated to Utah's Salt Lake Valley in 1847. To survive the wilderness, they quickly began diverting creeks that flowed from the Wasatch Mountains, using the water for crops. This was first accomplished at City Creek in Salt Lake City and quickly spread along the Wasatch Front. The Mormon's strong social mission, which tied the settlers together, emphasized the communal nature of their water projects—not unlike what had been present in the Spanish/Mexican/Indian

Southwest. The success of the Mormons in building irrigation-based communities inspired later American pioneers who settled in the West.

Texas was annexed by the United States in 1845; California and large swaths of the Southwest came under American sovereignty after the war with Mexico in 1848. The Treaty of Guadalupe Hidalgo, which ended the war, and the Gadsden Purchase Treaty in 1854 (where Mexico received $10 million for what are now parts of New Mexico and Arizona) promised that community water systems would remain. Acequias continued to be community-operated in New Mexico and were recognized as governmental units under state law (but with some limitations). By the end of the nineteenth century, many of these systems of public control—and the old standards of water rights—in the Southwest fell victim to a complex of new approaches shaped in large measure by local economics and political practices—emphasizing individual property rights.

During the California gold rush, the right to a land claim went to the first person working it. This "first in time, first in right" principle—or prior appropriation doctrine—could also apply to water. A miner did not acquire property in running water itself, but only its use if he continued to work the claim. The California State Legislature ultimately adopted the prior appropriation doctrine because it promoted economic development, but it gave no preference to communities over individuals. Eventually, every western state-endorsed some form of the doctrine, and nine states adopted it as its sole water law.

But the prior appropriation doctrine coexisted in California with riparian rights (all riparian landowners have an equal right to use the stream or river, which was the basic water law in the eastern United States) saddling the state with contradictory legal practices. Until 1886, California still struggled with having a dual system of water rights incorporating both prior appropriation and riparian rights. In that year, the California Supreme Court affirmed the dual system (*Lux v. Haggin*) in what became the "California Doctrine." Eventually, the California Doctrine also was adopted along the Pacific Coast (Washington and Oregon) and in the Great Plains (Nebraska, Oklahoma, Texas, Kansas, North Dakota, and South Dakota).

In the 1880s, Colorado invalidated riparian rights to surface water and began enforcing prior appropriation, which became the sole water right there. The so-called "Colorado Doctrine" came to dominate much of the Rocky Mountain region, including Utah, Wyoming, Arizona, New Mexico, Idaho, Montana, and Alaska. Wyoming, however, emphasized a different type of enforcement. The state constitution gave the state title to all water. The "Wyoming Doctrine" gave greater protection to appropriators than under the Colorado system. Nebraska, Oklahoma, North Dakota, and South Dakota also claimed full control over their water. No system worked without disputes or controversy, which also had proved the case in areas once under Spanish law.

The story of the building and maintaining of acequias in *Nueva Espana* extends the history of irrigation in the American Southwest and northern Mexico which had been going on for generations. The increased scale of the operations, the growth in water impounding capability, and the formality of

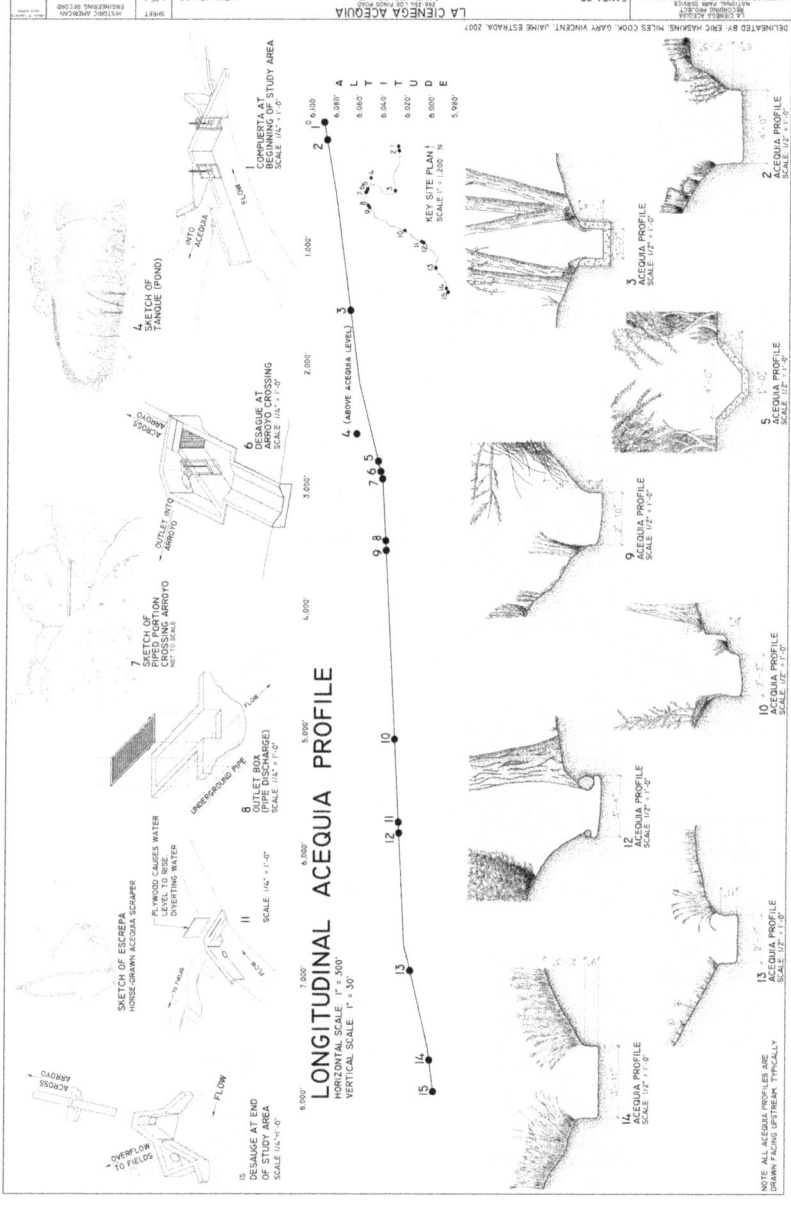

Figure 4.2 Longitudinal Acequia Profile – La Cienega Acequia, Los Pinos Road, Santa Fe, Santa Fe County, NM. Eric Haskins, creator. Public domain. Wikimedia.

the process through the Laws of the Indies demonstrate how natives and settlers in this arid, often parched region were able to establish a thriving agricultural system where there was none.

Most poignant for our examination of the environmental history of water in North America is how the acequias emphasized the importance of water as commons. Some have stated that acequias and its water symbolized "the spirit of a community." Along the upper Rio Grande and other locations in spring (at the start of irrigation season) there is an annual ritual cleaning extending from the acequia madre down to each field. Everyone is obliged to participate and celebrate.

In the most recent debates over privatization of water supplies, some critics have asserted the notion that commodifying fresh water is ethically wrong. Treating water simply as a product, they contend, leads to choosing the most profitable markets for providing water service, leaving some areas—especially poor communities and those located on the urban margins--without adequate service. The concern is that water is being treated more as an economic as opposed to a social and environmental good or "a shared resource and a public trust." In November 2002, the United Nations Committee on Economic, Social, and Cultural Rights adopted "General Comment No. 15" on the right to water. In referring to articles 11 and 12 of the International Covenant on Economic, Social, and Cultural Rights stated that "the human right to water is indispensable for leading a life in human dignity. It is a prerequisite for the realization of other human rights."

The recognition of the acequia and its water as a communal good, however, appears to be based largely on a practical footing. Indeed, the Moors considered water sacred and later generations of Spaniards, Mexicans, Pueblo, and others understood water as a precious resource. It gained such status in a hostile physical environment which required sufficient water for survival.

The best use of that resource was for it to be held communally, where everyone participated in collecting water and utilizing it as efficiently as possible. There were many disputes and disagreements over its control and use, but they did not diminish the necessity for linking water and community. Conditions in other parts of the world or in later times in the American West, however, did not necessarily lead to such an approach that proved so successful in the Southwest and northern Mexico for many years. But even there, conditions and context changed over time which ultimately limited water as a community-held resource, changing the role of access to water in fundamental ways.

Notes and Further Reading

Brown, John R. (2004) "Governing New Mexico's Water: Lessons from the Commons," International Association for the Study of the Commons," online, https://dlc.dlib.indiana.edu/dlc/bitstream/handle/10535/1878/Brown_Governing_040524_Paper498.pdf?sequence=1&isAllowed=y.

Fernald, Alexander G., Terrell T. Baker, and Steven J. Guldan (2007) "Hydrologic, Riparian, and Agroecosystem Functions of Traditional *Acequia* Irrigation Systems," *Journal of Sustainable Agriculture* 30: 147–171.

Gonzales, Moises (2014) "The Evolution of the Urban Acequia Landscape of the American Southwest," in C. Sanchis-Ibor, et al, eds, *Irrigation, Society, Landscape: Tribute to Thomas F. Glick* (Valencia: Universitat Poltitecnica de Valencia), 894–909.

Isenberg, Andrew C., ed. (2014) *Oxford Handbook of Environmental History* (New York: Oxford University Press), 513–552.

Meyer, Michael C. (1984) *Water in the Hispanic Southwest: A Social and Legal History, 1550–1850* (Tucson: University of Arizona Press).

Rivera, Jose A. (1998) *Acequia Culture: Water, Land, and Community in the Southwest* (Albuquerque: University of New Mexico Press).

Rivera, Jose A., and Luis Pablo Martinez (2009) "Acequia Culture: Historic Irrigated Landscapes of New Mexico," *Agricultura, Sociedad y Desarrollo* 6 (September-December): 311–330.

5 The Origins of Commercial Fishing in Newfoundland

Fishing is as old as civilization. Along with hunting and food gathering, it was a primary means of getting sustenance. Wherever there was a substantial amount of salt and fresh water there was likely to be fish and fishing. According to historian Poul Holm, "Fishing played a central role in discovery, and in the spread of humans throughout the globe" (Holm, 2004, 529). The search for fish as an important source of protein and for other uses was among the factors that drove early human exploration. Fish were relatively easy to preserve, transport, buy, and prepare. As we have seen in the earlier essays, water was an essential resource in human settlement and could offer a cornucopia of other resources—in this case abundance of fish on and around the coastal waters of Newfoundland. European colonialization of what became Canada depended on fishing for economic livelihood and even for political status and power.

Precontact settlements along the coastlines around the world took advantage of the abundance of marine life for survival and trade. Sea-hunting and fishing voyages beyond the coastlines prompted extensive travel for some peoples. Studies show that Japanese fishers by 2000 BCE sailed far into the Pacific, and Polynesian migrations likely were related to fishing. Greek merchants carried out extensive fish trade from the Black Sea and Russian rivers to Greek and Roman markets. In ancient Asia and Europe, aquaculture practices resulted in building fishponds for harvesting fresh fish for the wealthy. Fish as a source of food likely was as important or more important than meat in ancient Mediterranean cultures.

Improved curing practices and marketing helped further develop demand for fish by the Middle Ages. Emerging new governments also regarded maritime activities, including fishing, as good training for seamen in their navies. During the "Age of Exploration" (early fifteenth century through the seventeenth century), European fishers embarked further and further into deep ocean waters. By the end of the fifteenth century, more Europeans were involved in fishing than any other occupation save agriculture. The Portuguese sought tuna along the North African coast. Flemish and later Dutch fishers exploited herring in the North Sea. The English traveled to Iceland for codfish. While the fishing industry in Europe experienced ups and downs during this period, opening of vast cod fisheries in North America made

DOI: 10.4324/9781003041627-8

English, French, and Spanish fishing expeditions rich. In turn, the early fishery based on cod ultimately expanded to include more extensive fishing of other species and wider areas of fishing in the northwest Atlantic. (Cod fishing also took place in New England, Russia, and the northern coast of Norway).

Italian navigator Giacomo Caboto (known also as John Cabot) is credited with the re-discovery of Newfoundland in 1497. There he encountered an extensive cod fishery. The schools were so thick, he claimed, that a boat could hardly row through them. First Nations and Inuit had been fishing for thousands of years before this siting in what is now Canada—including Newfoundland—using nets, hooks, longlines, spears, and traps. The Beothuk (called Skraelings by the Vikings), native inhabitants on the east coast at the time of European contact, were not known to have fished offshore, however. As European interest in fishing in North America accelerated, some indigenous peoples throughout the Atlantic coastal areas faced a new level of competition for fishing sites, began trading practices with the newcomers, or suffered hostility and expulsion.

Information from early Viking voyages to Labrador (Helluland), Newfoundland (Markland), and the coast of Massachusetts (Vinland) led some European navigators to fish in the sea in this area during the fifteenth century before Caboto's voyage. But not until his observance of the fishing grounds did word spread quickly in Europe about the seemingly endless abundance of cod. Within ten years of Caboto's voyage, several European expeditions made annual trips to Newfoundland. The salt-cod commercial fishing industry became a bastion of Newfoundland's and Labrador's economy until the end of the nineteenth century. It took place primarily inshore off Newfoundland's coast but also included a Labrador fishery and extensive offshore bank fishing.

The Grand Banks were a major focus of European fishing in Canadian and adjacent international waters. The Grand Banks is a shallow area of submerged highlands extending over 36,000 square miles southeast of Newfoundland and east of the Laurentian Channel on the North American continental shelf. It consists of several separate banks including Grand, Green, and St. Pierre. (Banks whose tops rise close to the ocean surface can be hazardous to shipping and are referred to as *shoals*). They range from about 80–330 feet in depth. Some include the submarine plateaus that extend southwestward to Georges Bank (east-southeast of Cape Cod, Massachusetts). Here the cold Labrador Current meets the warm water of the Gulf Stream. This intersection plus the shape of the ocean bottom brought nutrients to the surface, creating conditions that produced one of the finest fishing grounds in the world. Marine life flourished there, but the environment was treacherous for fishers due to the unpredictable waves, heavy fog, occasional icebergs, hurricanes and other storms, and earthquakes.

The abundance of marine life in the Grand Banks thrived because of its shallowness and extensive range. It proved to be a good place for spawning, nursery, and feeding for a wide variety of fish and shellfish and supported large colonies of seabirds. Aside from the prized Atlantic cod, the Grand Banks was home to haddock, mackerel, capelin, halibut, redfish, turbot, yellowtail, witch flounder, American plaice, crab, lobster, scallop, and shrimp. Sea mammals such as seals, dolphins, and whales also were found there.

Until the late-twentieth century, cod were plentiful along the east coast of Canada and the United States, rich in protein, and easy to catch. They became one of the most commercially important fish as a source of food and cod-liver oil. For many years they were a mainstay in the diet of Europeans who fished relentlessly along the Grand Banks. Cod gathered in large numbers in the winter to spawn, which led to intense seasonal fishing (normally from March to October). Female codfish produced from 2 to 11 million eggs per year (some say 4–7 million), thus assuring large populations. The hatchlings drifted with the ocean currents for about ten weeks, and adult cod lived near the sea bottom. Cod are aggressive predators, eating capelin and herring as well as starfish, worms, squid, and their own young. They also undertook long migrations to shallow areas such as banks and bays.

The size of the cod population was dependent on several things: The eggs of older female codfish stood a greater chance of hatching than those of younger females, thus age distribution of the cod was important. Herring or other sea creatures ate massive amounts of fertilized eggs, which was not devastating if the overall cod population remained large. Once cod populations begin to decline significantly, fertilization dropped off dramatically and thus egg production plummeted. Overfishing or rock hopper dredging (which devastated sea bottoms where young cod took refuge) were the quickest ways to deplete the cod population, which became severe in the Grand Banks in the 1970s and 1980s.

In the fifteenth and sixteenth centuries, the quest for cod led to fierce competition among Europeans. In the early decades of the sixteenth century, Bretons and Normans from France, Portuguese, and Basques dominated European fishing activity in what is now Canadian waters. They fished and dried cod on the south and east coasts of Newfoundland and along the Strait of Belle Isle. About 1504 Basques, established seasonal whaling stations on the Labrador coast. By the 1520s, between 60 and 90 ships from French ports traveled to Canadian waters annually. Vessels—including those from New England—worked the coastal waters in Nova Scotia and the Bay of Fundy (south of Newfoundland). German settlers in Lunenburg developed a fishery largely through joint-stock companies. The Maritimes (the provinces of New Brunswick, Nova Scotia, and Prince Edward Island) built up their own large fishing fleets.

By the end of the sixteenth century, virtually all ports from Bristol to Cadiz participated in the fishing industry developed in North America. Early in the seventeenth century, the governments of France and England promoted the establishment of permanent settlements there, and their ships began to spend the winters in Newfoundland. French residents, however, were forced to leave the island when France ceded Canada to England in 1763 following the French and Indian War. By 1815, English-speaking people in Newfoundland largely replaced English migrant fishers inshore, and schooners based in New England and Newfoundland competed with European vessels in the Grand Banks.

By the nineteenth century, European ships fishing the Grand Banks were largely French, while Brazil and the Caribbean became major markets for cod. By the late 1820s, the Labrador fishery was firmly established. The Convention of 1818 partly resolved fishing conflicts between the United States and England after the War of 1812 and further were improved through a reciprocity treaty at mid-century.

The English fishers were few in number before 1570, but eventually controlled large areas of the eastern coast of the Avalon Peninsula (southeast portion of Newfoundland) by 1600. The slow start may be attributable to England's focus on its Icelandic fishery. Its vessels were based in seaports on England's northeastern coast on the North Sea, which was not well-located to take advantage of the Newfoundland fishing grounds. Also, vast markets for fish were not yet developed. Early on, Portuguese and Spanish fishers benefitted from the fact that they were staunchly Catholic with a strong demand for fish consumed on fast days. Also, French Atlantic seaports took advantage of the New World resource, became skilled at the cod-fishing trade, and developed inland markets in northern and eastern France.

Southwestern ports in England (in the "West Country") eventually proved better suited for such exploitation than their northern counterparts and would come to dominate the Newfoundland fishery, especially after Spanish and Portuguese fisheries faded. Decline took place in the sixteenth century when new taxes and greater governmental restrictions on trade and shipping weakened their ability to compete with their European rivals. (Portugal had been absorbed by Spain in 1580). Worst yet was England's war with Spain beginning in 1580, in which the English seized and destroyed many Iberian fishing ships.

France also benefitted from the decline of the Spanish and Portuguese fishing industries. Between about 1545 and the mid-1560s, the number of fishing vessels coming out of Bordeaux increased from 20 per year to 40; 12 to over 40 at La Rochelle; about 12 to 90 at Rouen. More substantial increases took place in the late sixteenth century in Les Sables d'Olonne, La Rochelle, and Le Croisic. While the English fishing fleet expanded from around 30 per year to as many as 200 by the end of the sixteenth century, French fishing vessels outnumbered English ships by about two to one in the mid-seventeenth century.

With the Europeans came fishing techniques used in the northeastern Atlantic during the Middle Ages, newer techniques developed later, ships and navigational principles used especially by the Portuguese and Basques in the fifteenth century, and modern mercantile capitalism. For example, in the 1800s many vessels replaced simple hooks and lines with longlines, which employed hundreds of hooks on groundlines set on the sea bottom. On offshore banks, larger schooners launched smaller "dories" to tend to the lines.

From the 1860s steam, vessels were equipped for seal hunting. By the end of the century, the "cod trap" (a cork-and-twine structure) was used to catch great volumes of inshore fish. New England fishers developed the *purse seine* for open water, which surrounded surface-schooling fish with a net hanging from a line of corks. The fishers tightened a purse line at the bottom of the net thus trapping the fish. All of these changes were meant to substantially increase hauls—and profits.

FISHING FOR COD ON THE BANKS FROM A DORIS.

Figure 5.1 Cod Fishing on the Banks from a dory, 1895. Unknown painter. Public domain. Wikimedia.

Aside from its great numbers, cod was the primary player in the exploitation of marine life because it could be dried or salted, could be transported long distances, and would keep for several months. In English, fisheries cod were caught close to shore in small boats, and then unloaded directly onto a *stage* or wharf where they were cleaned, split, and lightly salted. They then were dried on *flakes* (open tables). These shore-based *dry* (or *dry cure*) fisheries produced *hard-cure* cod that were prepared over the summer and then shipped back to Europe at summer's end. Although the salt was used sparingly, the product was referred to as *saltfish* and could be preserved for more than three years. Saltfish was favored in Portugal, Spain, Italy, Provence in France, and the Levant as well as in various interior areas of Europe. Such an enterprise led to a string of English and Irish settlements along the Atlantic coast and was the basis of England's territorial claims to Newfoundland.

French fishers not only operated along the shore but on the Grand Banks and other banks to a greater degree than the English. They had better access to salt than the English and most often processed their catch aboard ship. In the *wet* or *green* cure (*morue verte*—pioneered around 1550) cod were salted much more heavily than in the dry cure (or were packed in brine) and then were immediately shipped home. Thus, the French brought the fish to market faster than the English and allowed their vessels to return to the banks more than one time each season. The French lost a competitive edge when the English pushed them out of the Avalon Peninsula to Placentia (a little farther north). The French gave up claims to Newfoundland and Nova Scotia in 1713.

COD-FISHING ON THE NEWFOUNDLAND BANK.

Figure 5.2 Cod Fishing on the Newfoundland Bank, 1876. Public Domain. Wikimedia.

Beginning in about the second half of the nineteenth-century catches of cod increased greatly due to the introduction of trawling technology, but ultimately forced the inshore fishery to decline. Other technologies such as rock hopper dredges and electronic scanners to locate schools of cod were employed in the late-twentieth century. In the Maritimes after World War I, salted groundfish was a leading product, but fresh-fish trade became more important than before. In the 1920s, Newfoundland lost some of its markets to European suppliers but was able to compete more strongly in the West Indies (traditionally supplied by the Maritimes).

During the Great Depression, the Atlantic fisheries were hit hard. Fishing was a major employer, and thus the economic turndown badly affected myriad communities. After World War II, fishing fleets adopted new technology—including radios, radar, sonar, nylon nets, and various hydraulic gear—which helped them track and catch more fish, and also helped transport them over long distances through refrigeration. The federal government gave subsidies to fishers willing to build new vessels. In the 1950s unemployment insurance was extended to self-employed fishers, and the government also set up loan- and vessel-insurance programs. Significantly, various governments in the area encouraged the taking of new commercial species, including redfish, flounder, crab, shrimp, and scallops.

During the late 1960s, as much as 3.7 million tons of cod were caught annually. Overfishing led fish-producing nations to nationalize cod resources causing "cod wars." In Canada, cod fishing nearly collapsed in the 1970s. (In 1968 the cod catch from the Grand Banks was 810,000 tons; in 1974, 34,000 tons). A 200-mile national zone was established along the North American Atlantic coastline in 1977 to protect cod fisheries and limit conflict. Still, by the mid-1880s, Canada led the world in fish exports.

In the early 1990s, cod stocks dropped dramatically, leading to a moratorium along the Grand Banks and surrounding area off the coast of Newfoundland. Overfishing also took place in New England waters. Along with overfishing, poor regulations and climate change led to a major drop in cod catches by the late 1990s. Only more recently have the numbers grown, and more attention has been given to cultivating cod through aquaculture.

Fishing around Newfoundland and its environs was a central feature of European colonial rivalries in Canadian Atlantic region. Fish meant livelihood, wealth, and power. In time, the industry building around the Grand Banks and neighboring areas had worldwide economic implications. Cod, in particular, were sought after as a major source of protein. On land, water had been channeled through canals and irrigation systems to develop broad-scale agriculture. In the maritime world, water provided a different type of food source no less dependent on specialized technologies. That water as a source of food—and something to be vied for—was indispensable in both worlds, but also subject to misuse be it through aggressive depletion or overfishing.

Notes and Further Reading

Canadian Council of Professional Fish Harvesters (2021) "History of Fishing in Canada," accessed November 3, online, http://www.fishharvesterspecheurs.ca/fishing-industry/history.

De Loture, Robert (1957) *History of the Great Fishery of Newfoundland* (Washington D.C.: Gallimard, April).

Gough, Joseph, and Erin James-Abra (2015) "History of Commercial Fisheries," July 23, *The Canadian Encyclopedia*, online, https://thecanadianencyclopedia.ca/en/article/history-of-commercial-fisheries.

Holm, Poul (2004) "Fishing," in Shepard Krech III, J.R. McNeill, and Carolyn Merchant, eds., *Encyclopedia of World Environmental History* v. 2 (New York: Routledge), 529–535.

Kurlansky, Mark (1997) *Cod: A Biography of the Fish That Changed the World* (New York: Walker).

Lajus, Julia, and Dimitry Lajus (2004) "Cod," in Shepard Krech III, J.R. McNeill, and Carolyn Merchant, eds., *Encyclopedia of World Environmental History* v. 2 (New York: Routledge), 240–242.

Lear, W.H. (1998) "History of Fisheries in the Northwest Atlantic: The 500-Year Perspective," *Journal of Northwest Atlantic Fishery Science* 23: 41–73.

6 From Waterwheels to Steam Engines

According to the renowned Czech-Canadian scholar Vaclav Smil, "Energy is the only universal currency: one of its many forms must be transformed to get anything done" (Smil, 2017, 1). In the preindustrial period (and even extending into the early industrial age), four types of energy dominated the scene: human and animal labor, wind, biomass (organic material used as fuel), and moving water. Waterwheels became the most efficient converters of energy in the wood-based era and the early industrial period, opening new possibilities in a variety of fields from agriculture to mining. Ultimately, steam engines offered more power, encouraged a larger scale of production, and provided a wider variety of uses.

Waterwheels depended on moving water and steam engines on available clean freshwater (plus an external generator of heat) to function. Because they were stationary, waterwheels were directly linked to certain types of rivers and streams, while steam engines could be more selective in accessing water and fuel. Both energy sources took us far away from reliance on human and animal power on a grand scale. And both came with an environmental price—altering the landscape, escorting us into an industrial age with advantages and pitfalls (especially pollution), and reinforcing the tilted notion than humans—whenever and however—could alter nature to suit their needs.

Going back to antiquity, gravity irrigation and aqueducts were among the first kinds of water-based technologies. Observers asserted, however, that waterpower through waterwheels led to a major technological revolution in energy transmission across the globe. Gulzarilal Nanda, an Indian prime minister in the 1960s opined, "Water gushing out of thousands of springs at different places cannot move the wheels of a big engine to carry out very heavy tasks. But the channeled flow of the same water in the bed of a stream will, however, be irresistible and can become a source of tremendous energy."

Waterwheels were one of the earliest kinds of machine to utilize non-human, renewable energy. Traditionally, the mechanical power of flowing or falling water was converted into rotary motion by waterwheels. In the early nineteenth century, it was converted by water turbines which were used since the 1880s to drive generators to produce electricity. The first reference to the waterwheel was found in the writings of Antipater of Thessalonica in the first

DOI: 10.4324/9781003041627-9

Figure 6.1 Landscape with Waterwheel and Boy Fishing. George Caleb Bingham, 1853. Public domain. Wikimedia.

century BCE concerning grain milling. The origins of waterwheels, however, predate this mention by hundreds of years or more.

The horizontal wheel (known as a Greek or Norse wheel) had vanes extending from a wooden rotor, with a jet of water turning a rotor. The horizontal wheel turned millstones directly. The vertical wheel was more powerful and more efficient in either undershot or overshot designs. The undershot is a paddle wheel that turned under the impact of water current. It required gears to drive a millstone. Levels of rivers in the dry season could diminish the flow of water, and thus hamper the work of the undershot wheel. The overshot wheel, which required gears, received water from above, adding the impetus of gravity to the current.

The first description of a vertical waterwheel came from the Roman engineer Vitruvius (31 BCE–14 CE), although the first vertical waterwheels originated somewhere in the Middle East in the first century BCE. An important Roman application of the waterwheel took place at Barbegal near Arles in southern Gaul (France) in about 300 CE. The factory there, fed by Roman aqueducts, was a flour mill with 16 overshot wheels arranged in pairs on eight terraces. Flour produced in the mill could supply as many as 80,000 people.

Waterpower was important in ancient China for operating bellows to cast iron and for grinding grain. Chinese waterwheels typically were horizontal, but they also used industrial vertical waterwheels as early as 100 CE. Throughout the first thirteen centuries Chinese technology moved slowly from the East (carried first through central Asia) to the West along the legendary Silk Road. In medieval Europe, labor shortages among other reasons, encouraged waterwheel building. France led in the application of waterpower for industry in this period. In 1086, the *Domesday Book* (a manuscript record of the "Great Survey" of major parts of England and Wales ordered by William the Conqueror) listed more than 5,600 mills in southern and eastern England alone. Tidal mills apparently were an invention from the Middle Ages in both France and England. The Cistercian monastic order (formed in 1098) organized amazingly efficient factories using waterwheels. Increasing demand for metal in the Middle Ages led to miners using waterwheels to pump water from the mines, grind ore, run bellows at blast furnaces, and operate hammers at forges.

The builders of waterwheels were referred to as millwrights. They combined the skills of engineers and craftsmen. They decided where to dam streams using stone, wood, and dirt, and directed the water flow and fall. From a technical perspective, the waterwheel became an important link between artisanal work and future industrial processes. Historian of technology Robert Friedel has argued that "Perhaps the best illustration of the intensity of water installation in a single location and the technical impetus that it could provide came…in the southern city of Toulouse, on the Garonne" (Friedel, 2007, 37). By the twelfth century, at least sixty floating mills produced flour and meal for the surrounding area. The floating mills were protected by three dams, and eventually were combined into a stationary structure.

The largest of the dams was the Bazacle—probably the largest dam in the world at the time. As Friedel added, "The dams at Toulouse were particularly spectacular examples of one of the Middle Ages' most important contributions to power technology: the hydropower dam. The primary technical significance of the power dam lay in its making large and variable streams into useful sources of power" (Friedel, 2007, 38). Between the building of such dams and the refinement of the waterwheel, the Middle Ages offered technologies that help underpin the coming Industrial Revolution which began in Europe in the eighteenth century.

Indeed, waterwheels reached their highest status in the early decades of industrialization. But finding new waterwheels sites for the expanding textile industry in Europe had become more difficult because many worthwhile mill sites on usable rivers already were developed. At this time employing steam power—which was not so site-dependent as waterpower—was twice as expensive. The situation required, among other things, improving waterwheel technology such as iron-built wheels and large industrial waterwheels. In this early stage of industrialization watermills and horse-powered mills were central to cotton manufacturing. The mechanical machine for cotton clothing—the spinning jenny (1767)—was water-powered after 1769. In metallurgy, waterpower became the primary source for draining mines and smelting and fabricating metals.

In areas where steam power was particularly expensive, early industrial water turbines were employed. An invention of the early 1830s, a water turbine was a rotary machine that converted kinetic (energy in motion) and potential energy into mechanical work. Later in that decade two industrial sites in the Black Forrest in Germany were equipped with water turbines.

In America, waterpower technology was inherited from Europe. Small watermills were relatively easy to build, simple to maintain, and cheap to run. Historian David E. Nye has made clear, "In the first half of the nineteenth century, many Americans viewed mills not merely as transformers of raw materials and local economies but as the seeds of new communities" (Nye, 2003, 91). Mills remade places previously not developed for other purposes, or anchored agricultural locations.

The growth of agriculture also depended on the watermill to provide enough energy to transform crops into products for the community and the marketplace: corn into meal, wheat into flour, ore into metals, trees into lumber. In 1800, a typical small mill produced 15–20 horsepower, compared to a person turning a crank or on a treadmill generating only one-tenth of a horsepower (Nye, 1998, 22). Sawmills, more plentiful in the United States than Europe, not only enhanced town building but also attracted settlers and prompted the use of rivers for transportation of goods. The first American sawmill was built in 1611 in Virginia by German immigrants.

The watermill was not a perfect nor an inobtrusive technology. Requiring flowing or falling water, they were—at first at least—located in the countryside away from most large markets. They could not function in cold winter months nor during droughts. The wooden moving parts of the mill lacked strength and could warp or shrink. Noise from the large wheels and shafts was deafening. Watermills also competed for space and water with other activities such as fishing and agriculture.

Under normal conditions, a stream was dammed with a portion of the water diverted to the side through a millrace. The race was dug so that it was higher than the stream itself. The mill then was constructed and equipped with a waterwheel. Early on large rivers—such as the Merrimack, Connecticut, Hudson, and Schuylkill in the East—were too large to be dammed and "harnessed," but smaller streams and rivers saw the building of numerous mills throughout the nineteenth century. Nye has noted that along the Charles River in the Boston area mills were built in fishing grounds used by Native Americans and colonial settlers for catching shad, alewife, cod, bass, and smelt and for gathering clams, mussel, oysters, and eel along the shoreline. By 1812, there were 23 mills on the Charles producing flour, textiles, paper, dyes, and a variety of metal products (Nye, 1998, 23).

Fresh water, unlike land, was public as opposed to private property (access varied of course). It was open to common use requiring special treatment; water usage also was subject to unique practices imbedded in the law. Mills and dams for the first time raised legal questions over the relationship between property law and private development.

Clashes between riparian users were typical, however, and the watermill inevitably came into conflict with other stream uses. Aside from the waterwheel, the dam was the most essential element of a mill. The dam created a storage reservoir, or millpond, which not only obstructed navigation and log floats but also blocked the seasonal movement of fish. In some cases mills were required to provide fish ladders or fishways to safeguard migration of several species, but enforcement was spotty at best.

Since water was an important source of energy, contested uses of water became the focus of voluminous litigation over rights. Problems of pollution from sawdust produce by sawmills and the dumping of dyes, acids and other waste into rivers also drew lawsuits. As mills expanded in scale, water pollution or diminishing of water quality became increasingly contentious. These problems were acknowledged, for example, in Massachusetts in the late-nineteenth century. Although an 1878 act prohibiting the discharge of human waste, refuse, or other pollutants into watercourses within twenty miles of a public water supply failed to make the law retroactive. Industrial rivers like the Merrimack, Connecticut, and parts of the Concord river near Lowell were exempt from the legislation. Economics over nature.

The productive capacity of the mills, particularly flour and lumber mills, gave them substantial privileges under the law. The Mill Act (1713) in Massachusetts allowed mill owners the right to build dams even if upstream riparian owners objected to possible flooding or impact on their lands. Numerous state laws came to support the building of mills because they represented economic progress. As historian Ted Steinberg observed about the role of water in the coming industrial age in New England,

> The law provided the framework for settling conflicts over water. But that framework was evolving in a way that reduced water, more and more, to an abstract commodity...By the latter part of the nineteenth century it was commonly assumed, even expected, that water should be tapped, controlled, and dominated in the name of progress—a view clearly reflected in the law.
>
> (Steinberg, 1991, 16)

Labor disputes grew as larger commercial mills expanded operations into the early industrial era. In the 1840s, women went on strike at the Lowell mills seeking shorter hours. The mills often were freezing in the winter and stifling hot in the summer. Women in the textile industry often worked 70–75 hours for a six-week period, and frequently by candlelight. This was a class issue as well because the workers occupied the lower rungs of the social ladder in their communities, receiving meager wages for their demanding work. Protesters likened the mills to prisons. Supporters of slavery in the South strongly criticized the free labor in the cotton mills, arguing that mill workers were not as well cared for as black slaves. Child labor was a serious problem as well, even though some states tried to curb it. Labor unions arose to take on the poor conditions in the mills, the long hours, and general worker exploitation, but pitiable labor practices continued into the latter years of the industrial revolution with little substantive change.

The importance of watermills to the United States economy and their impact on the watercourses where they were located was never clearer than with the acceleration of the Industrial Revolution in the latter nineteenth century. In essence, the energy from falling or moving water drove industrial development until steam power overtook that role. Mills grew in size and number as the demand for energy increased. In some respects, controlling water for milling helped improve the skills necessary for constructing canals and locks in the coming "Canal Age" stimulated by the building of the Erie Canal.

Large factories increasingly were being built near or in some modest-sized and small Eastern cities by the 1820s where fast-running river water was most accessible, but this excluded major cities such as New York City or Philadelphia. The mills in Lowell, Massachusetts (on the Merrimack River) were the major centers for textile and machinery manufacturing, followed by those in Manchester and Nashua, New Hampshire; Lawrence, Chicopee, and Holyoke, Massachusetts; and Biddeford and Saco, Maine. The Merrimack River—the "hardest working river in the world"—runs through New Hampshire and Massachusetts and became the archetype industrial river in the United States. It provided approximately 80,000 horsepower to 900 mills and factories in 1880. Lowell's industrial capacity exceeded any water-powered cotton mills in the country. According to Smil, "…more water power capacity [in Europe and the United States] was added during the first six decades of the nineteenth century than ever before, and most of these machines continued to operate even as steam power, and later electricity, were conquering the prime mover markets" (Smil, 2017, 155).

Cotton cloth and woolen goods proved to be among the first mass-produced products in the country and helped change patterns of consumption which had been dominated by home-made products. Mass-produced goods began to gain a competitive advantage that made new investors and producers rich. The South, however, lagged behind the budding industrial water-powered North in modernizing its economy. Slave labor and a less energy-intensive type of industrial development clearly distinguished it from its northern neighbors.

The role that waterpower played in early industrialization both in Europe and the United States often has been obscured by the role of steam power. In many traditional historical interpretations, waterpower essentially was considered a local energy source that rapidly declined in importance in the early industrial revolution as more modern sources came to the fore. Nineteenth-century English writer George Eliot (Mary Ann Evans) had an interesting view of the steam engine: "Ingenious philosophers tell you, perhaps, that the great work of the steam-engine is to create leisure for mankind. Do not believe them; it only creates a vacuum for eager thought to rush in."

Figure 6.2 First US steam fire engine, 1840. Unknown author. Public domain. Wikimedia.

Improvements in stationary waterpower in the nineteenth century were often overshadowed by the application of steam power as an energy source for riverboats and locomotives. The successful adaptation of steam power to transportation did not lead to its universal adoption in all other sectors. Not until the 1880s did steam exceed waterpower as an energy source for industrial expansion in the United States. Furthermore, the use of steam power was highly concentrated in only a few industries, such as saw-milling. Since the cost of steam power initially was higher than for waterpower, industries used it only when they needed freedom of location away from viable waterways. Stationary steam power was above all an urban phenomenon, where ready access to waterpower was absent but interest in manufacturing great.

In time, steam power overtook waterpower as an industrial energy source, especially with the adaption of abundant and cheap coal to steam engines. As Smil asserted, the steam engine was "the first practical, economic, and reliable converter of coal's chemical energy into mechanical energy, the first inanimate prime mover energized by fossil fuel" and had "a profound importance for global industrialization, urbanization, and transportation..." (Smil, 2017, 235) Yet the application of the steam engine to industrial processes was slow even with coal. In 1820, there were about 100 waterwheels for each steam engine in use at the time.

In a steam engine, hot steam—supplied by a boiler—expanded under pressure. Part of the heat energy was converted into work. For the best efficiency, the remainder either could vent or could be condensed in a separate apparatus (condenser). Steam could push a piston (connected to a crank on a flywheel to produce rotary motion) in a reciprocating engine, or the steam could be discharged at high velocity through a series of blades in a steam turbine.

Commercialization and widespread use of steam engines took more than a century to develop. In Great Britain, the use of steam power in draining mines existed 100 years before its broad industrial applications. Even during more rapid diffusion after 1820 it continued to compete with waterwheels and water turbines. Its full impact did not take place until after 1840 with the construction of railroads and steamships and greater use in manufacturing.

The steam engine's development had a long history, but its use as a practical device began with French physicist Denis Papin's experiments with small models in 1690. English inventor and military engineer Thomas Savery developed a small, steam-driven pump (operating without a piston) in 1698. By 1712, English blacksmith and inventor Thomas Newcomen had built a steam engine (using atmospheric pressure) to power mine pumps, and his invention began to spread in England by 1750. Scottish inventor and mechanical engineer James Watt received a patent for his improved steam engine and condenser in 1769. The Watt engine—the dominant design for all modern steam engines—quickly became a commercial success in coal mining, manufacturing, and transportation. Watt's largest units matched the most powerful waterwheels, plus they had mobility which waterwheels could not claim.

In the textile mills of the northeastern United States, steam-powered machinery came into use as a supplement to waterpower especially during the dry season when water flow declined. Some fifteen years after the construction of the Lowell mills more than 100 steam cotton mills were operating in the United States. After the Civil War even more coal-fired steam boilers/machines replaced water-powered machines. From 1869 to 1889, steam power represented 96 percent of the increase in horsepower available to manufacturers. Even Lowell mills shifted to steam: between 1867 and 1880 the number of steam engines more than doubled there (from 31 to 73). After 1900, waterpower got a modest boost because of the development of hydroelectric facilities and the transmission of electricity to factories, commercial establishments, and cities, but it eventually faced more competition, even in the production of electricity.

The urbanization of manufacturing in the late-nineteenth century, the boom in railroad construction, and general population growth all pushed steam power forward. Technical improvements in steam engines also narrowed the gap in cost between waterpower and steam power—along with the wide availability of coal.

The need for motive power was central to the success of steam engines. European countries with their numerous roads and canals and slow-paced rivers had little need for the steamboat—but this was not true in the U.S. Steamboats came of age in North America after the War of 1812. By 1830, they dominated internal transportation throughout the U.S. The New York to Philadelphia route was the most important traffic lane in the East. In the West, the steamboat was vital to commerce and travel. By 1855 there were 727 steamboats operating in the West compared with only 17 in 1817. The abundance of wood was a major factor in the geographic distribution of steamboats. Acquiring supplies of wood was not a problem in the West, but in the 1840s, steamboats in the East began to turn to coal because of diminishing supplies of wood, easy access to cheap coal, and the superior burning capability of this fossil fuel.

As crucial as it was to the economy of the American West, the steamboat was limited to navigable rivers. Railroads offered greater flexibility at a reasonable cost. By the 1870s, trains were the last word in American transportation— already spanning the continent. By 1840 there was only 1,818 miles of track in all of Europe, but more than 3,000 miles in the United States (almost 9,000 by the 1850s), which was twenty-nine years before the completion of the first transcontinental railroad in 1869. The combination of improved track building (especially after 1869) and the perfection of motive steam power gave the railroads their competitive advantage.

The use of steamboats declined steadily in the late nineteenth century, but railroads continued to thrive as the dominant transportation medium until the advent of the automobile. Trains also outlived their dependence on wood. The adaptation of coal to the steam engine made the relationship between coal and transportation essential. Anthracite (or hard coal), which was highly concentrated in northeastern Pennsylvania, was used in steam engines as early as the mid-1820s and became a practical source of steam power by 1830. By 1835, steam-driven factory machinery, steamboats, and locomotives were using anthracite. The mutual dependence of anthracite and the railroads— with railroads monopolizing most of the mines—changed the very nature of the coal industry in the United States.

The replacement of anthracite coal with bituminous (or soft) coal began in the late nineteenth century. Unlike anthracite, bituminous coal was mined in at least twenty-two states (90 percent in Pennsylvania, Ohio, Illinois, Virginia, and Maryland), and not as prone to monopoly. By 1880 coal—mostly bituminous—was used for about 90 percent of locomotive fuel. Coal also constituted the largest single item of revenue tonnage for the railroads to haul.

The widespread use of bituminous coal (and to a lesser extent anthracite), including use by steam engines of all types, had one serious drawback—highly toxic smoke. Increased consumption of bituminous coal from the 1870s through World War I contributed greatly to pervasive pollution problems especially in industrial cities. The use of wood had led to deforestation and its own air pollution in cities; anthracite mining in northeastern Pennsylvania scarred the countryside. Air pollution from anthracite was bad, but from bituminous much worse. An irate New Yorker called for the abolition of anthracite furnaces because they "will hourly destroy the health of our women and children." The negative effects of coal smoke became particularly obvious in cities during the 1890s. There were few scientific measurements for smoke pollution then, but annoyances to the sense were enough to set off protests. A physician writing in 1915 stated, "The old idea that black smoke is indicative of a prosperous community is still too prevalent…What we look upon as normal may be wholly abnormal; conditions need correcting." In the 1910s almost every city had smoke-abatement leagues bent on initiating tougher antismoke laws (Melosi, 1985, 32–33).

Smoke pollution was not a national issue at the time, but closely linked to cities like Pittsburgh, Cincinnati, St. Louis, and Chicago, where soft coal fired steel mills, other industrial plants, and railroads. While smoke was viewed as a visible sign of prosperity and material progress it was a stigma for these cities and others until they turned to other fuel sources in the 1920s.

For those of us living in the twenty-first century, it is difficult to imagine a pre-petroleum, pre-electrical power energy regime like the one that existed in the eighteenth and nineteenth centuries in the United States. The relationship between water and energy production was never as great. The internal combustion engine that power automobiles, trucks and other conveyances was a product of the twentieth century as was the electrification of homes and factories. Without waterpower before then, human and animal power—plus wind power and biomass of course—would have to carry the load so to speak. Long-distant and large-scale transportation depended not only on watercourses, but on water for steam engines to power steamboats and locomotives—and for a very short period of time some automobiles.

The energy of flowing water and the tides made early industrial production possible, with steam engines expanding upon that auspicious start. Technical advances using water ushered in breathtaking economic success, but also altered the landscape, sometimes fouled the air and water, and encouraged exploiting human labor despite greater expectations for mechanization and automation. In all of this, good and bad, water was the lynchpin, an indispensable resource in a very different way than it had been in the world prior to the eighteenth and nineteenth centuries.

Notes and Further Reading

Friedel, Robert (2007) *A Culture of Improvement: Technology and the Western Millennium* (Cambridge, MA: MIT Press).

Hunter, Louis C. (1979) *A History of Industrial Power in the United States, 1780–1930, Vol. 1 Waterpower in the Century of the Steam Engine* (Charlottesville, VA: University of Virginia Press).

Jones, Christopher F. (2014) *Routes of Power: Energy and Modern America* (Cambridge, MA: Harvard University Press).

Melosi, Martin V. (1985) *Coping with Abundance: Energy and Environment in Industrial America* (New York: Knopf).

Nye, David E. (2003) *America as Second Creation: Technology and Narratives of New Beginnings* (Cambridge, MA: MIT Press).

Nye, David E. (1998) *Consuming Power: A Social History of American Energies* (Cambridge, MA: MIT Press, 1998).

Penna, Anthony N. (2020) *A History of Energy Flows: From Human Labor to Renewable Power* (New York: Routledge).

Smil, Vaclav (2017) *Energy and Civilization: A History* (Cambridge, MA: MIT Press).

Steinberg, Theodore (1991) *Nature Incorporated: Industrialization and the Waters of New England* (Amherst, MA: University of Massachusetts Press).

Part III

Expansionism and Western Settlement

In this section, we explore two important water episodes that grew out of nineteenth-century United States expansion West. The first treats the California Gold Rush emphasizing how the use of hydraulic mining turned individuals' scramble for riches into a systematic corporate enterprise and its implications. The second—the use of the Rio Grande River as a boundary between Mexico and the United States after the Mexico–U.S. war of 1848—visits the problematic nature of using water to separate political entities with less regard for the implications in doing so.

Traditional studies of Western expansionism stress the success that the United States achieved in creating a two-ocean empire. Myriad histories discuss pioneers on the Oregon Trail, the Louisiana Purchase, war with Mexico, the independence of Texas, the Oregon controversy, and the annexation of Florida. Recent histories treat more emphatically what impact those events had on the people—especially Native Americans and Mexicans —that were defeated, conquered, killed, or displaced in order to secure the new southern western, and northern borders of the United States. There also is substantial discussion, especially among environmental historians, about the aridity of the West and what that meant for establishing and sustaining life on the Great Plains and beyond. Studies of westward expansion range anywhere from Frederick Jackson Turner's "The Significance of the Frontier in American History" (1893) to Richard White's *Railroaded: The Transcontinentals and the Making of Modern America* (2012).

Yet water played decisive roles in a variety of enterprises not often represented in the big issues of exploitation and expansion westward. Water was no less significant, however, in raising our awareness of changes coming to the transcontinental United States and also to our neighbors—in this case—to the south. The need for water, and its exploitation in several different ways, was historically important.

Large-scale gold mining was not possible without hydraulic techniques and the siphoning of millions of gallons of water from rivers and streams adjacent to the mining sites. For some, the use of this method made them rich, but at the expense of small-scale placer miners. Hydraulic mining was a form of business consolidation, where capital investment in a new technology paid

DOI: 10.4324/9781003041627-10

great monetary dividends. Such an approach to mining was made possible by weak—or sometimes promotional—laws that gave greater leeway to mining companies than to individual miners. Furthermore, environmental protection of rivers, streams, forests, and fauna was ignored in the name of short-term economic gain. Water became a driver of the new technology, while watercourses suffered its repercussions. It was ironic that in a part of the world where water was so vital for living, agriculture, and urban growth, so much of it was squandered to recover a shiny metal.

In the case of the Rio Grande, its role in the dual history of Mexico and the United States extends over many generations. Beyond its value as an indispensable source for farming and agriculture, it became—through war and politics—a physical boundary separating two nations. That moving water is less enduring as a line of demarcation than even a simple set of markers placed in the dirt and confirmed by surveyors, courted many problems between landowners and governments. Pushing the boundary of Texas south to the Rio Grande had a tumultuous history, but once the dividing line was established, the nature of the river itself did not support the objective of distinguishing Mexico from the U.S. Complicating matters was that the Rio Grande served other purposes than just a boundary. It was a source of water for many activities, a location for human settlement and businesses, a haven for many plants and wildlife, and most recently a battleground over a wall to keep people from moving north.

Climate change poses yet another variable in understanding the Rio Grande's identity. In a Fall 2015 story, "Climate Change on the Rio Grande," on the World Wildlife Fund (WWF) website, Dr. J. Alfredo Rodríguez-Pineda (hydrologist for WWF-Mexico) stated about lands around the river,

> We expect less frequent, more intense rains with climate change. We are going to have heavy rains, maybe in shorter bursts. When that happens, the water has no time to infiltrate the ground, so the soil starts flowing away with the water. Soil erosion is to nature like cancer is for humans. It destroys little by little. And farmers have been unintentionally destroying all this landscape little by little by applying the wrong farm techniques.

This watercourse, and others, has many uses, many problems, and many meanings.

7   The California Gold Rush: Placer and Hydraulic Mining

January 1848 was auspicious for what became the largest gold rush in American history starting in the Sierra Nevada, a mountain range between the Central Valley of California and the Great Basin. On the 24th day of that month, gold was discovered at John Sutter's mill in Coloma, El Dorado County, California. By the next year word of the strike occasioned a glut of nearly 100,000 prospectors from around the world to seek their fortune in California.

Water became a formidable tool in extracting gold. Hydraulic mining was a much more efficient, albeit wasteful and environmentally destructive, technique than panning for gold in a river or stream. The story of the California Gold Rush and its impact on the region, its people, and its water is often camouflaged by the vivid image of the Forty-Niners (and the Forty-Eighters just before them) who became a centerpiece of California history and American folklore.

The California Gold Rush had several long-term impacts. As historical geographer Lary M. Dilsaver stated, "The effects of the rush were crucial to later western development: the rise of San Francisco as a financial center, the expansion of mineral-related settlement throughout the cordilleran West, and the development of agriculture in fertile valleys" (Dilsaver, 1985, 1). But the dominance of mining in the Sierra Nevada foothills where the strike began also hindered development of other land uses, and ruined watercourses and major animal habitats. More drastic, however, was that the gold rush led directly to the destruction of local Native American settlements and—significantly—to the dispersal or outright slaughter of several tribes. This result was little short of genocide.

The impact of wide-ranging new settlement and uncontrolled extractive mining also was devastating to the environment, especially rivers and streams. Using new techniques, especially hydraulic mining, watercourses and surrounding natural areas were severely disrupted including the damage to redwood and other forests.

Rivers were the first place that gold was found and exploited through placer (alluvial) mining. Later hydraulic techniques demanded vast amounts of

DOI: 10.4324/9781003041627-11

water–often acquired through impounding streams and dam building—ravaged hillsides. The results of such activity silted and polluted many rivers, despoiling them for generations. Water does not necessarily first come to mind when we think about the legendary mining of gold in California, but it is at the heart of the story.

The earliest documented discovery of gold in the United States took place in 1799 on the Little Meadow Creek in Cabarrus County, North Carolina. The strike led to further exploration and mining in the southern Piedmont area (Virginia to Alabama) from the 1830s until the Civil War. Mining resumed after the war but peaked by 1887. In 1829, another gold rush took place in Georgia. Thousands of prospectors stormed the foothills in the southern Appalachian Mountains, where unruly mining towns popped up, millions of dollars in gold were extracted, and the Cherokee were pushed out of their homeland. Historians Robert V. Hine and John Mack Faragher asserted, "What happened in Georgia was a preview in miniature of what would happen in the Golden State, as well as dozens of other mining strikes throughout the West" (Hine and Faragher, 2000, 236).

California was part of a large geographic area where gold deposits led to a mad rush for wealth and prosperity. Molten gold formed in the Sierras (and also in the Rockies and Cascades) during the prehistoric uplifting of the mountains. It poured into fissures or veins, and the veins located close together formed a lode that could be tapped at or near the surface. Placers were places where gold eroded into small particles through weathering and the action of water.

The California Gold Rush was concentrated in three major regions: the first discoveries occurred along the American River and other tributaries of the Sacramento River; then the tributaries of the San Joaquin River (which flowed north joining the Sacramento in the delta east of San Francisco Bay); and also in the Northern and Southern Mines along the Mokelumne River. Other strikes took place to the northwest around the Trinity, Klamath, and Salmon rivers. Gold also was discovered in Southern California in 1842, but on a smaller scale than in Northern California six years later.

The mining activities went through boom–and–bust cycles between 1848 and 1855. About 1,400 towns or camps sprung up in four counties in the Sierra Nevada area: Nevada, Placer, El Dorado, and Amador. The fate of the mining towns varied. Yet, between 1849 and 1858, California mining produced over $550 million in gold. In the 1850s that represented one-third of the world's gold production.

Figure 7.1 Print shows hydraulic mining for gold in California. Published in *The Century Illustrated Monthly Magazine*, January 1883. Henry Sandham, *The Monitor*. Public domain. Wikimedia.

Part of the reason why news of the strike in 1848 spread like fever was because early prospectors were able to simply pick up gold nuggets and flakes from the streams. These opportunities did not last long, and the miners turned to placer mining. Running water in a river or stream led to deposits of gold forming in the riverbed, on riverbanks, or in inundated areas along the watercourse. These were known as gold placer deposits and had been mined for many centuries. Unlike hard rock mining, which extracted veins of minerals from solid rock, placer mining separated minerals like gold from sand, eroded rock, or gravel. The simplest method was for a miner to use a prospector's pan to collect the heavier gold from the lighter rubble scooped from the water. This was back-breaking work and of limited scale, since a miner washed about fifty pans in a usual twelve-hour workday to obtain just a small amount of gold flakes or dust.

To accelerate the process and to cover more area some prospectors turned to utilizing human-made flumes (small aqueducts or channels) to guide the water through wooden troughs (sluice boxes, cradles, or rockers). In the bottom of the trough or rocker were riffles (little cleats) which agitated the slurry of water and gravel allowing small particles of gold to fall out because they were heavier than the sand or gravel. Some miners added small amounts of mercury to the bottom of the trough or rocker (used in greater abundance as mining in-dustrialized), since it could trap particles of gold. Heating the mercury released the gold. Another form of gold placer deposits was buried under debris or solid rock and required different mining techniques.

The image of a grizzled, old white prospector swishing a gold pan in a river has been part of gold rush lore for decades. The image is a stereotype that does not bear up to reality since miners in nineteenth-century California could be Native American, Mexican, Canadian, Chinese, Hawaiian, African, African American, Australian, Chilean, Peruvian, all kinds of European, and sometimes women as well as men.

California's indigenous people participated in the gold rush as miners. One government report stated that in 1848, more than half of all prospectors were Native American—sometimes referred to as "diggers." Historian Andrew Shaler tells the story about Edward Gould Buffum, a Euro-American gold-seeker who led a small party to the slopes of the Sierra Nevada Mountains toward the Yuba River. The party had no knowledge of where to find gold but happened upon a village of Nisenan people. Buffum was surprised to learn that a group of Nisenan traveled to the Yuba River to mine gold daily. They would exchange their gold for flour, meat, and other supplies in nearby towns while another group hunted game (Shaler, 2020, 79–80).

During the early stages of the gold rush, before it turned frantic, stories about the Nisenan were common. Some of the Native American miners worked for white landowners, but many more mined independently. Decades before the gold rush Indian labor had been the basis for the regional economy in California. The Hispanic labor system that had exploited Native Americans now was moving from the ranchos to mining.

Maidu and Nisenan peoples who resided in the northern Sierra foothills were among the first to mine for gold on their ancestral lands. James Marshall long has been credited with discovering gold at Sutter's Mill in January 1848, but historian James J. Rawls argued that "it is not unreasonable to suggest that a California Indian should be given credit for that momentous discovery" since Native Americans led the Marshall party to the mill site and had been working there previously (Rawls, 1976, 29–30). Indigenous peoples from outside California also joined the gold rush. Mexicans were among the first migrants to seek mining opportunities, joined by Yaqui people from Sonora. Cherokee and Wyandot also arrived from the Great Plains to seek their fortune.

More than one hundred different Native American tribes, constituting about 300,000 people, resided in California before contact. By about 1870 only 30,000 remained; in 1900 only 16,000 survived. About 60 percent died of new diseases from settlers. The remainder were killed at the hands of incoming migrants seeking gold—supported by government sanction. What amounted to genocide was accompanied by unspeakable cruelty—scalps sold for 25 cents and severed heads going for $5. Thousands of native children were sold as slaves.

By the end of 1849, with thousands of outsiders pouring into California, there was a dramatic rise in violence against Native Americans. In two years, non-native population soared from 15,000 to over 165,00; by 1860, 400,000. At best these indigenous people were seen as cheap labor, at worst savages who were an obstacle to seeking gold. In the 1850s, there were very few Native American miners. According to Shaler, "A party of Euro-Americans from

Oregon Territory, among the first overland immigrants to reach California, arrived in Nisenan territory in 1849. Violence erupted after the Oregonian men invaded a Nisenan village and raped its women. The Nisenan retaliated and killed five of the Oregonians, commencing a cycle of violence that would lead to catastrophe for Nisenan people" (Shaler, 2020, 81).

State-sanctioned volunteer militia campaigns against the Nisenan occurred in 1850 and 1851. Diggers were cheated out of their mines. In 1850, the California State Legislature passed the Act for the Government and Protection of Indians, which denied Native Americans the right to testify in court and gave the right to white Americans to keep them as indentured servants. Local communities compensated those who killed the natives. The federal and state governments alike condoned all manner of behavior among whites to deal with the "Indian Problem" in order to make room for new settlers and to lay claim to gold on tribal lands. In 1850, approximately 400 Pomo people—including women and children—were massacred by the U.S. Cavalry and local volunteers at Clear Lake north of San Francisco. This was one example of many.

Intolerance was widespread during the gold rush. While no group faced the horror of extermination like Native Americans, foreign non-whites faced nativism, bigotry, and racism common at the time. They most often were seen as interlopers and competitors in the gold fields. Chinese immigrants (many from Guangdong province) entered California first in small numbers after 1848, but most of them who mined did so as other prospectors began to abandon placer claims. The Chinese became the main placer miners in the state from 1852 to 1853. Mexicans faced discrimination as did those seeking gold who were of African descent (free blacks, slaves, and people from the Caribbean and Latin America). In 1850, the California legislature passed the Foreign Miners' Tax Act imposing a levy of $20 per month on foreign miners. Criticism of the act and its failure to raise enough revenue resulted in the act being repealed in 1851, and subsequently replaced by the Foreign Miners' License Tax Act of 1852, charging $3 per month. The state also levied a "commutation tax," which required shipmasters to submit a list of all foreign passengers and to post bond for them once the ship made port.

Women faced discrimination as well. Especially early in the gold rush, there were few white women and children in the boom towns. Men always were seeking "feminized service industries" as historian Albert L. Hurtado stated, such as laundresses, cooks, housekeepers, seamstresses—and prostitutes (Hurtado, 1999, 6). Under the best of circumstances, white women could seek marriage in this gender-imbalanced society. But opportunities beyond the domestic life were scarce. Some women mined, with poorer women tried their hands at gold panning to make a decent living. Women of color sometimes faced harsh treatment and even violence, especially if they were Native American or Mexican. As was typical of the era, women were forced to adapt to social and economic conditions that they did not establish.

The change in mining techniques in the 1850s also influenced who became active in mining, who dominated the enterprise, settlement patterns, and how the mining-impacted California. The decline of placer mining along the western slope of the Sierra Nevada began in 1854. Chinese prospectors, blocked from other mining operations, continued to mine the tailings, and some farmers engaged in seasonal placer mining, but this labor-intensive approach never really recovered on the scale of the late 1840s and early 1850s. As placer mining became less profitable and was being replaced by larger-scale practices such as hydraulic mining, the impact on the physical environment proved to be much greater and longer-lasting than experienced in the first two or three years of the gold rush.

Quartz (hard rock) mining was an alternative to placer mining. Miners began exploiting outcrops of gold-bearing rocks that they located upstream (through gravel ridges) from where they had been panning. They sunk shafts as far down as 300 feet and then crushed the ore. The technique came from Nevada, where it had limited success in the Comstock. Through a variety of technological improvements (and the infusion of more capital from Nevada in particular), they were able to sink shafts to more than 1,200 feet and substantially increase the return from the ore. Quartz mining was active in a narrow band from Mariposa County to northern El Dorado County (the Mother Lode) with more than 300 mines. Richer veins were located around Grass Valley and Nevada City to the north. Corporate mines controlled the best locations.

Hydraulic mining was an alternative to quartz mining, focusing on utilizing high-pressure water in lower-grade gravel and eroded rock to extract gold. It could accumulate larger amounts more easily, more economically, and in greater quantity than panning or using a sluice box or sluice boxes. It also extended the life of the placer mines.

A form of hydraulic mining had been used by the Romans on Spanish placer deposits two thousand years earlier. The newer technique first employed in California was capital-intensive industrial mining as opposed to the labor-intensive panning and sluicing process. Hydraulic mining would dominate mining in the West for a century. It relied on expensive tools, grander-scale engineering techniques, and cheaper labor—converting small owners into wage earners. And demanded lots of water.

Hydraulic mining (hydraulicking) used a powerful stream of water to extract gold from mine tailings, placer deposits, alluvium (deposits left by flowing streams), laterites (soil rich in iron oxide), and saprolites (soil rich in clay). In California, natural banks and cliffs in the gravel along the sides of gulches were an early source for exploitation. The high-pressure stream of water broke up the material and suspended it in a slurry. The slurry was moved into sluices where the gold settled behind baffles with the lighter material washed away.

The water used in the process was routed through a nearby river, stream, or dam into ditches or flumes to a point upslope from the area to be worked. At the mine site, water was collected in a tank or vertical shaft to build pressure and then run through a hose to a nozzle or monitor. The monitors, which replaced the nozzles, were huge water cannons—some up to 18 feet in length—spraying enormous amounts of water under great pressure that could wash away mounds of material down to bedrock. Some miners claimed that the monitor could throw a stream of water 400 feet into the air.

The use of this method originated in 1852–1853. In April 1852, Frenchman Antoine Chabot—who had a gold claim on Buckeye Hill near Nevada City—ran a canvas hose into a water ditch on the hill above him. This let the force of the water wash gravel into the sluice which he could loosen with hand tools. Ground-sluicing, as it was called, made it possible for miners to wash large amounts of gravel every day. The next year, Edward Mattison (sometimes spelled Matteson) and Eli Miller worked with Chabot to make a larger canvas hose. Tinsmith Miller replaced the original carved wooden nozzle with a tapered one made of sheet iron. (Later, canvas hoses were replaced by leather, rubber, crinoline, or sheet iron.) Mattison and Miller directed the stream of water against the bank and realized they had developed a tool to replace the back-breaking work of shoveling gravel and rock with something more efficient (Kelley, 1954, 345–346).

Several companies designed and made end attachments with names such as *Hoskin's Dictator, Hoskin's Little Giant,* or the Craig Company's the *Monitor* (which became the common use). The monitor was from one to eight inches in diameter and capable of washing down a whole hillside. It was estimated that an eight-inch monitor could spew 185,000 cubic feet of water in an hour with a velocity of 150 feet per second (or 100 miles per hour). The stream could carry a fifty-pound boulder, and records indicate that men were killed by the force of the water from 200 feet away.

Some monitors operated twenty-four hours a day and demanded vast amounts of water almost beyond measure. The most active region of hydraulic mining was on the San Juan Ridge of the Sierra Nevada between the middle and south forks of the Yuba River. Companies flocked to the area. At the North Bloomfield mine (located on a 3,300-foot ridge) sixty million gallons were used each day. (Enough water to supply a city of 3 million.) The company president estimated that 16 billion gallons of water were used in 1876 alone. At Malakoff Diggins north of Nevada City—the world's largest hydraulic gold mining operation—eight giant cannons shot 100,000 gallons of water per minute all day, every day. In thirty years, 40 million cubic yards of earth were excavated at Malakoff leaving an open pit of more than a mile in length and 600 feet deep. Hydraulic mining soon spread to Colorado, Idaho, Montana, South Dakota, and Nevada.

Acquiring and impounding water became essential to the new mining process. The Sierra Nevada annually had one of the heaviest snowfalls in the world (an average of ten feet was common) and thus the local rivers and streams carried large amounts of water in the spring and into the summer. Some years, however, droughts challenged the mining operations. Without an abundance of water, hydraulic mining was not possible.

The simplest method was placing a small dam across a stream to divert water. Soon gouging was utilized, which entailed digging a rough trench in which water could flow. More elaborate methods followed, including larger and longer sluices to clean the gravel in search of gold. Mining companies built extensive ditch and flume systems to capture the water. Eventually, large companies, like the Eureka Lake and Yuba Canal Company, built such systems. By 1882, there were 6,000 miles of flumes, ditches, and sluices in the Sierra Mountains. Many early ditches and reservoirs were slapdash and unreliable, causing serious accidents, and would have to be upgraded or rebuilt.

The environmental fallout from hydraulic mining was devastating. Preservationist John Muir observed that because of the practice "the hills have been cut and scalped and every gorge and gulch and broad valley have been fairly torn to pieces and disemboweled, expressing a fierce and desperate energy hard to understand" (Isenberg, 2005, 41). Miners were ignorant of the environmental damage, paid little attention to it, or defended their rights to use their property as they saw fit.

Not all of California's environmental problems began with the gold rush. By 1849, the new state-to-be experienced several changes because of human action (such as agriculture, lumbering, and hunting) including extensive open-range livestock grazing; introduction of exotic grasses from the Mediterranean region; destruction of some animal habitats; extensive hunting of grizzly bear, elk, pronghorn antelope, and many more animals; and poaching of sea otter and other marine mammals by Americans, Aleuts, and Russians.

Yet the gold rush accelerated the damage to grasslands, tule marshes, wildlife, forests, and other natural areas in substantial ways. In the case of forests, the mining industry demanded large amounts of wood for building canal systems and for fuel, thus hastening the clear-cutting of trees. (As population grew, the demand for lumber to build houses and other structures also increased dramatically). Thousands of acres of rich farmland were barren with orchards and grain fields buried by silt and mining waste.

Rivers, streams, and adjacent land in the mining regions experienced the greatest devastation. It took about two tons of rock to recover each ounce of gold. The powerful monitors washed away thousands of acres of soil from hillsides and forests silting up and choking downstream rivers and other watercourses. By doing so they also changed the course of rivers, destroyed fisheries, covered farmland with sand and gravel, obstructed navigation, triggered flooding, and caused serious erosion. Often great amounts of cyanide, mercury, and other toxic chemicals were spilled into the water. About 211 million cubic yards of residue entered the Yuba, American, and Bear rivers alone. Miners washed away an estimated 885 million cubic yards of debris into the Sierra Nevada valleys and ravines equal to 3½ times all the material excavated to build the Panama Canal or about 1,000 years of Mississippi sentiment loads.

Towns such as Marysville and Yuba City (at the confluence of the Yuba and Feather rivers) repeatedly flooded as the bed of the Sacramento River lost sixteen feet. Heavy rains in 1861–1862 resulted in the first severe flooding prompted by hydraulic mining creating extensive sheets of water and mud in the southern Sacramento Valley, including on the grounds of the State Capitol. On January 19, 1875, an extremely disastrous flood raced through the Sacramento Valley filing the streets of Marysville with thick, sticky mud essentially destroying the city. The high levees built around Marysville created a huge bowl for the floodwaters to enter, which not only ruined property but killed some residents. The State Engineer reported that North Bloomfield Gravel Mining alone had dumped more than 20 million cubic yards of dirt, rock, and debris into the Yuba. Ironically, Mark Twain had once referred to Marysville as "the most well-built city in California."

After the 1875 flood, Marysville residents formed the Anti-Debris Association, and similar committees from a variety of farming communities in the Central Valley were also formed to counteract the Hydraulic Miners' Association and its supporters (which included state political leaders). The industry had enjoyed prosperity especially between 1853 and 1859 but began a slow decline thereafter. The industry nevertheless continued to be buoyed up by the Mining Act of 1872, which gave the mining companies virtual free rein on public lands.

Hydraulic miners showed no interest in stopping their practices especially as opponents increasingly criticized them. As historian Duane A. Smith noted, "Almost from its beginnings, the industry was preparing a defense for hydraulicking, much earlier than for any other facet of mining" (Smith, 1987, 67). Making changes was too costly and would ruin the industry, they argued. There would be a great loss of jobs and towns would be destroyed. The miners defended their property and water rights (frequently a point of contention) touting ingenuinely that the rivers belonged to everyone. A couple of publications even promoted hydraulic mining as a tourist attraction—the great water arcs of the hydraulic canons, "the mountain removing to the sea" (Smith, 1987, 68).

Downstream farmers were tired of mud and silt inundating their orchards and croplands every spring and burying roads. The protests increasingly intensified in the 1870s, especially after the Marysville flood. Farmers found allies in railroad companies, especially the Central Pacific Railroad (later the Southern Pacific). This was a fascinating outcome given that miners depended so heavily on the railroads in their business. Railroads, however, wanted their riverside tracks to be mud-free, but most significantly they were concerned about the environmental damage because they were major property owners in the valley.

Litigation against the hydraulic miners started in the 1860s and 1870s. With the backing of the railroad, farmer groups turned to state and federal courts for a remedy to their problems. In 1882, New Yorker and Marysville-area landowner Edwards Woodruff filed a lawsuit against North Bloomfield and other mining companies. The suit requested a permanent injunction against further operations because of the negative impact on his property. To support Woodruff's plea, his legal defense pointed to a state law (April 1855) meant to protect owners of crops and related structures against damage from mining operations. The other side offered numerous pro-mining statutes. The legal battle raged.

The court rejected the miners' claim that Congress or the state had authorized the use of the river for deposit of debris. Miners countered by arguing, among other things, that they could not be solely responsible for filling the Sacramento and Feather rivers. In fact, they added, the hydraulic companies had acted in good faith in building dams and working their claims, and that the injunction would essentially kill the goose that laid the golden egg(s). On January 7, 1884, the judge (a Forty-Niner himself) rendered his famous 225-page decision in *Edwards Woodruff v. the North Bloomfield Gravel Mining Company* rejecting the defendant's claims and "perpetually enjoined and restrained" them from discharging or dumping debris into the Yuba River or its tributaries (Smith, 1987, 72).

Ultimately, hydraulic mining was superseded by a variety of other economic activities, including lumbering and farming. Congress passed the Caminetti Act in 1893 in an attempt to revitalize hydraulic mining. The law allowed hydraulicking if a catchment basin was used to trap debris. By this time, most companies were out of business and the expense of rebuilding was just too great. This created ghost towns that remained empty until the Great Depression.

After *Woodruff* mining companies converted the scooped-out basins of hydraulic mines into ponds and installed steam-powered dredges to excavate the gravel by the ton. Especially after the turn of the century, dredges were used for digging low-grade deposits in the flat valleys of the Mother Lode area. Iron dipper buckets shoveled up sand and gravel ran through a revolving cylinder and then dumped into a separator with angle irons (processed with mercury) to capture gold flakes. Larger rocks and gravel went out of the back of the dredge creating huge tailings. This proved to be a profitable enterprise but was devastating to the alluvial topsoil in the area. Because of unrelenting pressure, mining companies tried various methods of reclamation in the areas of their digging with varying degrees of success.

Without easily available water the whole hydraulic enterprise that marked the gold rush in the 1850s and beyond would not have existed. Environmental historian Andrew C. Isenberg rightly noted, "The riverine environments of the western slope of the Sierra were the ecological centers of the gold country. The riverlands were unique environments in the otherwise dry climate of the foothills and Central Valley." He added, "River flow alternately impeded and permitted miners' search for gold. Higher water in the spring prevented miners from scouring the auriferous gravel of the riverbeds for treasure. In the summer, when the streams had receded, companies of miners sought to dam or divert streams to search the riverbeds for dust, flakes, and nuggets" (Isenberg, 2005, 26–27).

Hydraulic mining exploited local water sources in a unique way—one that was intrusive and damaging. Without that abundant water derived from the snowpack (although sometimes capricious) and pushed through water cannons, the environmental story of this part of California would be quite different. Returning to Isenberg: "In effect, the operators of hydraulic mines profited by extracting valuable commodities from nature while passing along the costs of extraction to less powerful members of the community in the form of industrial pollution. The transformation of nature to make possible the extraction of gold was extensive, invasive, and, when one considers the damages to riverlands, forests, fisheries, and human health, arguably as costly as the wealth extracted from the mines was valuable" (Isenberg, 2005, 51). Water proved to be a commodity as precious as gold.

Notes and Further Reading

Brands, H.W. (2002) *The Age of Gold: The California Gold Rush and the New American Dream* (New York: Random House).

Chan, Sucheng (2000) "A People of Exceptional Character: Ethnic Diversity, Nativism, and Racism in the California Gold Rush," *California History* 79 (Summer): 44–85.

Dilsaver, Lary M. (1985) "After the Gold Rush," *Geographical Review* 75 (January): 1–18.

Haywood, Joseph J., Jr. (1981) *The California Debris Commission: A History* (Washington, DC.: U.S. Army Corps of Engineers).

Hine, Robert V., and John Mack Faragher (2000) *The American West: A New Interpretive History* (New Haven, CT: Yale University Press).

Hurtado, Albert J. (1999) "Sex, Gender, Culture, and a Great Event: The California Gold Rush," *Pacific Historical Review* 68 (February): 1–19.

Isenberg, Andrew C. (2005) *Mining California: An Ecological History* (New York: Hill and Wang, 2005).

LeCain, Timothy J. (2009) *Mass Destruction: The Men and Giant Mines that Wired America and Scarred the Planet* (New Brunswick, NJ: Rutgers University Press).

Kelley, Robert L. (1954) "Forgotten Giant: The Hydraulic Gold Mining Industry in California," *Pacific Historical Review* 23 (November): 343–356.

Rawls, James J. (1976) "Gold Diggers: Indian Miners in the California Gold Rush," *California Historical Quarterly* 55 (April 1): 28–45.

Shaler, Andrew (2020) "Indigenous Peoples and the California Gold Rush: Labour, Violence and Concentration in the Formation of a Settler Colonial State," *Postcolonial Studies* 23, 79–98.

Smith, Duane A. (1987) *Mining America: The Industry and the Environment, 1800–1980* (Lawrence, KS: University Press of Kansas).

Taniguchi, Nancy J. (2000) "Weaving a Different World: Women and the California Gold Rush," *California History* 79, 141–168.

8 Capricious Border: The Rio Grande River

Boundaries between cities, states, and countries can be physical and/or political. Some last for decades, others change more frequently. There can be hard borders and soft borders; invisible borders or quite visible borders; and some—under stressful circumstances—can disappear overnight. Water boundaries present unique challenges. Whereas natural obstructions to transportation by land can create physical boundaries, rivers can change course or dry up, then change course again. They possibly represent the weakest—and most unpredictable—separation between two political entities. The Rio Grande dividing parts of Mexico from the United States is one of those borders.

The two countries share a border of almost 2,000 miles—the world's single most crossed international boundary—made up of various landforms and the Colorado and Rio Grande Rivers. The latter is a major part of that boundary stretching about 1,254 miles. The dividing line on the Rio Grande is the deepest channel on the river. According to historian Jacqueline E. Timm, "...because of the meandering nature of the river, innumerable land changes have taken place [along the border], some cases arising from the river's shifting course. The many cases arising from the river's shifting course have included jurisdiction over San Elizario Island, Morteritos Island, and the bancos [large land protrusions] before 1905; cattle seizure; wing dam and fence construction; the Brownsville Wharf Case; water diversion by the American Rio Grande Land and Irrigation Company; and the Chamizal Dispute" (Timm, 2020, 1).

Sometimes flooding took its toll on altering the river. Other than environmental change or natural occurrences, disputes arose over badly drawn maps, political friction or rivalries, and a variety of legal claims. Some observers viewed the international boundary as unsatisfactory, others chronicled bitter and complicated relations between the two countries because of the nature of the border (made more difficult in recent years because of President Donald Trump's call for a permanent wall separating the countries). In all cases, a river boundary of any kind was likely to be unreliable. The history of the relationship between Mexico and the United States is connected to its water border.

DOI: 10.4324/9781003041627-12

Although a historically important river, the Rio Grande physically has not been and is not an impressive body of water. However, at 1,900 miles it is the fifth-longest river in North America and the twentieth longest in the world. In Spanish, the river is called *Rio Grande del Norte* and in Mexico, it is referred to as *Rio Bravo* (or *Rio Bravo del Norte*). It is not wide, often sluggish, and not particularly scenic for much of its length.

The Rio Grande originates as mountain streams at more than 12,000 feet in the San Juan Mountains (at the eastern face of the Continental Divide) in southern Colorado, flowing southward through New Mexico to El Paso, Texas. At that point, it becomes part of the international boundary between Mexico (states of Chihuahua, Coahuila, Nuevo Leon, and Tamaulipas) and the United States. From El Paso it flows toward the Gulf of Mexico through a subtropical coastal plain and into the Atlantic Ocean.

The Rio Grande Basin drains a more than 330,000-square-mile area in the southwestern United States and northern Mexico, which includes numerous vegetation types, desert and woodland ecotones (a region of transition be-tween two biological communities), and a variety of urban and rural peoples. The tributaries of the Rio Grande greatly expand its reach. They include the Pecos, Devils, Jemez, Chama, and Puerco Rivers in the United States and the Conchos, Salado, and San Juan Rivers in Mexico.

The Rio Grande has a long history of human habitation and use, with intermittent periods of conflict. Indigenous people lived close to the river for more than 10,000 years, some building rock shelters and growing crops such as corn. Pueblo Indians called the Rio Grande *Posoge* (*P'Osoge*) or "big river." Discovery and conquest of the region by European explorers began in the sixteenth century. Spanish explorer Alonso Alvarez encountered the river in 1519. At the time the river's mouth was wide enough for Alvarez to sail up it for twenty miles. In 1536, Spaniard Alvar Nunez Cabeza de Vaca was the first European to cross the Rio Grande upstream. That was the first time the "Rio Bravo" appeared on maps. The earliest settlements began in 1563 in Chihuahua.

In 1595, Don Juan de Onate was awarded a contract by King Philip II of Spain to settle what is now New Mexico. In 1598, he forded the Rio Grande at El Paso del Norte. According to geographer D. W. Meinig, "Onate pro-ceeded like many another predatory adventurer beyond the reach of official supervision, conquering [Pueblo], punishing, plundering whatever lay at hand, searching for mines and riches never to be found, settling into exact tribute from a badly disrupted and sullen society." Not until 1606 was he "recalled, fined, and exiled from his new province" (Meinig, 1986, v. 1, 16).

In the sixteenth century, various Spanish explorers observed irrigated sys-tems on the river. As we have observed earlier, they used the indigenous labor to develop large acequias to irrigate fields in the seventeenth century. From the sixteenth through the early nineteenth centuries several famous explorers found their way to the Rio Grande/Rio Bravo including Francisco Vasquez de Coronado (1540) and Zebulon Pike (1806).

Figure 8.1 Rio Grande watershed showing dams and diversions, October 7, 2012. Creative Commons Attribution-Share Alike 3.0. Wikimedia.

European conquest and state-building turned the Rio Grande River into a commodity to be fought over as a resource and as a boundary. Spanish/Mexican settlers established themselves in the northern reaches of *Nueva España* above the Rio Grande/Rio Bravo by the late sixteenth century. But villages and towns remained sparse along the whole river, sometimes subject to Comanche raids. By the mid-eighteenth century, New Spain relaxed efforts to colonize present-day Texas as protection against French incursions. The threat from the French was removed when France ceded the vast Louisiana territory to Spain in 1762. The secret treaty came at the end of the last battle of the French and Indian War (which confirmed British control of Canada) at a point where the French were exhausted from the long and expensive conflict.

France especially wanted to keep the immense tract of land from British hands. As a fading power, Spain returned the Louisiana territory to France in 1800 under pressure from Napoleon Bonaparte. He hoped to build an American empire centered upon the sugar trade in the Caribbean. Napoleon ultimately gave up on these ambitions and sold the territory (828,000 square miles) in 1803 to the United States for a pittance—$15,000,000. The ceded land stretched east to west from the Mississippi River to the Rocky Mountains.

The United States had bought the Louisiana territory with undefined boundaries, which also was true when France acquired it from Spain. President Thomas Jefferson wanted negotiations with Spain to fix the boundaries, arguing that the area not only encompassed the western Mississippi Valley but the Gulf Coast from the Rio Grande in the west to the Perdido River (in Florida) in the east. Spain believed the purchase was invalid and refuse to cede so much of its territory. The Spanish government had internal problems and suffered foreign invasion, and the United States simply annexed West Florida as far east as the Perdido River in 1810.

Under President James Madison, the United States prepared for renewed negotiations with Spain in 1816 to settle claims with respect to Florida, Texas, and the Pacific Northwest. The negotiations did not begin until 1818 when James Monroe was president and John Quincy Adams, Secretary of State. Adams and Spanish minister Luis de Onis completed the treaty (Adams-Onis Treaty or the Transcontinental Treaty) in 1819 with the United States receiving Florida and the boundary being fixed from the Sabine River to the Pacific Ocean. It produced the first solid American claim to lands on the Pacific Ocean but left the American claim to Texas very weak—essentially ceding it to Spain. The treaty assumed that the border between the United States and New Spain was the Sabine River, but some American settlers in Texas still claimed that it was the Rio Grande.

By 1821, Mexico had successfully declared independence from Spain, and refused to observe the treaty boundary line. Efforts at surveying the border failed, and the line remained unsurveyed until the early 1840s. Between 1819 and about 1836, the Nueces River (parallel to the Rio Grande and running fifty to 100 miles northeast through Corpus Christi) was proclaimed to be the boundary between Mexico and Texas, especially by the Mexicans. When Texas declared independence from Mexico in 1835, the Treaty of Velasco established the border along the Rio Grande. While President Antonio Lopez de Santa Anna signed the treaty, the Mexican Congress would not ratify it, and Mexican presidents after Santa Anna did not acknowledge Texas independence. Needless to say, the border between Texas and Mexico remained in limbo throughout the life of the Republic, with Texas adamantly claiming the Rio Grande as its rightful western and southern border.

The region between the Nueces and the Rio Grande had been called *La Colonia de Nuevo Santander* under Spanish rule and *Tamaulipas* under Mexico. Starting in 1767, the Spanish throne assigned tracts of land in the area, each having access to water mostly on the Rio Grande. Under Mexico, ranching was the principal economic activity there. As early as the 1820s American settlers slowly entered what became known as the *Trans-Nueces*. Conflict over the introduction of slavery increased tensions. "Throughout the years of the republic," Meinig stated, "the country between the

Nueces and the Rio Grande remained unsettled and unsecured from either side; maps of the time referred to the 'Mustang or Wild Horse Desert,' and it was known as a dangerous haunt of Indians, 'mustangers,' and 'prairie pirates'" (Meinig, 1993, v. 2, 141).

At the very least the dispute over whether the Nueces or the Rio Grande formed the border between Mexico and the Republic of Texas was a symptom of the growing tensions between Mexico and the United States. One month after the United States annexed Texas on December 29, 1945, President James K. Polk sent American troops to the Rio Grande. When Mexican forces crossed the river in April 1846, he declared war on May 13. The action of the Mexican army was just the pretext the expansionist president needed to claim that Mexico had invaded the United States. American troops pushed south as far as Mexico City, and Mexico surrendered on February 2, 1848.

In the Treaty of Guadalupe Hidalgo, Mexico was forced to give up nearly 55 percent of its territory (more than 525,000 square miles). In addition, the Rio Grande (between its mouth and El Paso in the west)—and not the Nueces—would form a major part of the border between the two countries. The parties agreed that surveyors from each country, working together, would map the new 2,000-mile border. However, the task proved more difficult than anyone imagined largely because of the difficulty of the work in the desert and in other rough terrains, fear of Indian raids, political infighting, and financial mismanagement. In addition, adequate mapping of the United States' southern border could not be completed until after the Treaty of Mesilla—or the Gadsden Purchase—in 1853 which reestablished the southern boundary of New Mexico and Arizona.

The Gadsden Purchase entailed 29,670 square miles of land south of the Gila River and west of the Rio Grande, where the United States was interested in building a transcontinental railroad line (later completed by the Southern Pacific Railroad in 1881–1883). The purchase also was intended to resolve other border issues. Santa Anna agreed to the sale for $10 million, money needed by his financially beleaguered Mexican government. Some believe Santa Anna was willing to make the deal rather than have the territory simply seized by the Americans. Historian Odie D. Faulk concluded that if the southern border of New Mexico had been set as the 1848 treaty as the makers intended, there would not have been need for the Gadsden Purchase and a southern transcontinental railroad route would have been already secured (Faulk, 1962, 226).

In 1855, Army surveyors from both countries completed a boundary survey which included the Rio Grande and provided climate, landscape, flora, fauna, and other details. Yet, the imprecision of setting a boundary along a natural feature like the Rio Grande insured no permanent solution to the border issue. Floods, falling banks, land loss, and more would make it difficult to fix the markers that defined the border. Historian Norris Hundley, Jr. suggested that part of the difficulty in establishing proper international law related to the boundary—aside from considerably different theories posed by the two parties—could be traced to the treaties negotiated after the Mexican-American War:

> The negotiators of the Guadalupe-Hidalgo treaty in 1848 and the Gadsden Purchase in 1853 took little note of the three river basins [Rio Grande, Colorado, and Tijuana] divided by the boundary they drew. Since settlement in the region was sparse, they gave no thought to the development of the basins or to any future conflicts that might arise over water… (Hundley, Jr., 1966, 18)

Beginning in the 1850s, the fickleness of the river—meanderings, floods, seasonal dry spots—made clear how unstable the new borderline would be. The population was growing, property owners complained, and public officials had few answers. Political scientists Alan C. Lamborn and Stephen P. Mumme suggested that significant territorial disputes between the countries occurred between 1852 and 1868 when the channel of the Rio Grande between El Paso and Ciudad Juarez shifted to the south. In places, the shift covered as much as 800 acres. The most serious shift occurred in 1864 when approximately 600 acres of land was displaced to the United States side of the border (Lamborn and Mumme, 1988, 52).

A convention held in 1884 was a first step in attempting to reach a satisfactory solution to boundary problems. The resulting treaty called for the dividing line between the two countries along the Rio Grande to "follow the center of the normal channel" irrespective of alterations of the banks through erosion or even if the original channel bed was dry or obstructed by deposits. In 1889, the International Boundary Commission was established with the responsibility for applying boundary and water treaties between Mexico and the United States, conducting surveys, monitoring boundary and river movements, and settling any differences that might arise. The commission initially was established as a temporary body but made permanent in 1900.

The 1889 Convention, for all its value, did not resolve another important naturally occurring problem along the Rio Grande—the formation of *bancos*. As the river snaked along the flat valley floor, it created wide loops and curves. When seasonal flows strengthened, direct routes could form between points on a curve. Thus, bancos formed in these cases. As historian Sarah Hill stated, the banco was "an island of land, formally belonging to one nation, but cut off from it by a change in the river's course and wholly contained in the other nation" (Hill, 2011, 903). In a 1905 convention, the two countries agreed that areas of bancos lying north or south of the river would fall under the respective jurisdiction of either the United States or Mexico (with some size and population exceptions). The treaty affected 215 bancos. Property holders, however, were not satisfied with such a simple solution and lawsuits persisted.

Aside from boundary issues, disputes arose about access to the water for irrigation and other purposes over several decades especially beginning in the late-nineteenth century as populations increased on both sides of the border. In the El Paso-Juarez area, Mexican farmers often lacked access to sufficient supplies of water, many having to abandon their farms as a result. Mexicans accused Texans of bringing on the shortages and vice versa. In several cases, up-river diversions had reduced the flow. A treaty in 1906 guaranteed Mexico at least 60,000 acre-feet of water per year from the Rio Grande (in case of drought the amount would be reduced) in exchange for Mexico relinquishing its claims to Rio Grande water between El Paso and Fort Quitman (about 75 miles south of El Paso). However, the 1906 treaty did not resolve some major problems including apportionment of water on the lower Rio Grande and on the Colorado River. Other water-supply disputes cropped up all along the border for years to come.

Probably the best-known border disagreement along the Rio Grande was referred to as the Chamizal dispute, which began in the late-nineteenth century. Spring floods by 1873 had moved the river almost one-half mile south of the *Paso del Norte* boundary, according to Hill, "leaving several hundred acres of what had been Mexican land on the north side of the river. With each new change in the river's course, Americans promptly claimed and began to subdivide and sell the land within the city of El Paso" (Hill, 2011, 902–903). In response, the government in Mexico filed suits on behalf of hereditary titleholders in Mexican courts in 1895. Americans refused to leave the land called the Chamizal tract (a sizeable area between El Paso and Juarez) despite favorable rulings in the Mexican courts for its citizens.

Not until 1911 did the International Boundary Commission convene in El Paso to attempt to settle the Chamizal dispute. They ruled in favor of the Mexican claim, arguing that most of the river movement occurred because of the major 1864 flood. Americans refused to accept the decision and continued to develop the land under dubious titles. The dispute did not end until the Chamizal Treaty was signed in 1963, which acknowledged the 1911 decision. Mexico received compensation in acreage in the Chamizal and east of Cordova Island. The river border along El Paso-Juarez is now lined with concrete so the river will not shift its channel.

Throughout the twentieth century, boundary disputes were increasingly linked to water rivalries and even "water wars" (and of course immigration) not easily resolvable by the boundary and water commission. In 1933, the two governments set up the Rio Grande Rectification Project to straighten and stabilize the river boundary through the growing El Paso-Juarez Valley, control flooding, improve irrigation and soil conservation, and divide water equitably between the two countries. It also called for the construction of a floodway and levees on about eighty-five miles of the river near El Paso and the construction of Caballo Dam to generate electricity. According to historian Joanne Tortorete Kropp, "The Rio Grande Rectification Project was a rare instance of bi-national cooperation in an otherwise acrimonious relationship between the United States and Mexico" (Kropp, 2016, vii).

The second water distribution treaty—the 1944 Water Treaty—between the United States and Mexico, authorized the distribution of waters on the Rio Grande from Fort Quitman to the Gulf of Mexico (and on Colorado), and construction, operation, and maintenance of dams on the main channel. Through the treaty, the International Boundary Commission became the International Boundary and Water Commission—the IBWC—known in Mexico as the *Comision Internacional de Limites y Aguas* (CILA). It expanded the old commission's responsibilities, especially taking on more issues related to water rights and conflicts, questions of sanitation, and problems related to salinity. Over decades the IBWC built concrete channels for parts of the river (to prevent course changes) and constructed dams.

While settling some long-standing border issues, the efforts of the IBWC and related actions agreed upon by Mexico and the United States, natural occurrences such as drought and human intervention in the controlling of the Rio Grande had serious environmental implications (as well as economic repercussions). Efforts to straighten the channels, which reduced the length of the river, often increased flooding. Dam building changed the river's hydrology and contributed to silting. These are problems that developed along other rivers as well where "harnessing nature" was the goal, but also proved to create many problems. Irrigation, agriculture, and industrialization (see the chapter on the maquiladoras) added many polluting chemicals to the river. Controlling water flow and high demand for water on both sides had an impact on flora and fauna (Wiedenfeld, 2011, 1125–1126). For example, of the 45 fish species in the mid-portion of the Rio Grande in New Mexico only 17 are native (two are now extinct) (Baker Jr. et al., 2004, 201). The debate over Trump's wall in recent years further aggravated concerns about the river's flow, its ecology, and the tensions it might cause with Mexico.

Trying to control the river boundary for the purpose of limiting or curtailing animosity between the United States and Mexico did not always achieve its goal. Disputes over who is getting or not getting a "fair share" of the water, for example, go on to this very day. Unanticipated consequences that disturbed the ecology, threatened economic development and had an impact on human settlement appeared and reappeared over the years. In many respects, the Rio Grande served as a tangible dividing line, although regularly contested, between the two nations. The physical realities of a water border, however, were variable and seemed to change often. The case of the Rio Grande stood to prove that water boundaries are erratic and have political, social, and environmental consequences.

The role of the Rio Grande in the history of Mexico and the United States went beyond its status as a boundary. The river was a resource (water, fuelwood, fish, and game), a political and social dividing line (especially the incendiary immigration crisis), an icon, and an entity influencing international law. Water has many uses and historical meanings along the Rio Grande.

Notes and Further Reading

Baker, Jr., Malchus B., et al, eds. (2004) *Riparian Areas of the Southwestern United States: Hydrology, Ecology, and Management* (Boca Raton, FLA: Lewis Publishers).

Faulk, Odie B. (1962) "The Controversial Boundary Survey and the Gadsden Treaty," *Arizona and the West* 4 (Autumn): 201–226.

Gelbach, Frederick R. (1993) *Mountain Islands and Desert Seas: A Natural History of the U.S.-Mexican Borderlands* (College Station, TX: Texas A&M University Press).

Hill, Sarah (2011) "Mexico and the United States," in Kathleen Brosnan, ed., *Encyclopedia of American Environmental History* v. III (New York: Facts on File), 902–907.

Horgan, Paul (1984) *Great River: The Rio Grande in North American History* (Hanover: Wesleyan University Press).

Hundley, Jr., Norris (1966) *Dividing the Waters: A Century of Controversy Between the United States and Mexico* (Berkeley, CA: University of California Press).

Kropp, Joanne Tortorete (2016) "Constructing a River, Building a Border: An Environmental History of Irrigation, Water Law, State Formation, and the Rio Grande Rectification Project in the El Paso-Juarez Valley" (Dissertation, University of Texas, El Paso, January 1).

Lamborn, Alan C., and Stephen P. Mumme, (1988) *Statecraft, Domestic Politics, and Foreign Policy Making: The El Chamizal Dispute* (Boulder, CO: Westview Press).

Meinig, D.W. (1986) *The Shaping of America: A Geographical Perspective on 500 Years of History, Volume 1: Atlantic America, 1492–1800* (New Haven, CT: Yale University Press).

Meinig, D.W. (1993) *The Shaping of America: A Geographical Perspective on 500 Years of History, Volume 2: Continental America, 1800–1867* (New Haven, CT: Yale University Press).

Scurlock, Dan (1998) *From the Rio to the Sierra: An Environmental History of the Middle Rio Grand Basin* (Fort Collins, CO: Forest Service, United States Department of Agriculture).

Timm, Jacqueline E. (2020) "Rio Grande Boundary," *Handbook of Texas*, online, https://www.tshaonline.org/handbook/entries/rio-grande-boundary.

Wiedenfeld, Melissa G. (2011) "Rio Grande," in Kathleen Brosnan, ed., *Encyclopedia of American Environmental History* v. IV (New York: Facts on File), 1123–1126.

Part IV

Commerce, Industry, and Urban Growth

The nineteenth-century brought the United States and Canada to modernity in special ways. It was a period of nation-building, figuratively, materially, and culturally. Elsewhere, I have referenced historians Peter N. Stearns and John H. Hinshaw who stated, "Industrial revolutions constitute those rare occasions in world history when the human species alters its framework of existence" (Stearns and Hinshaw, 1996, vii). Thus, an industrial revolution is not just an economic event, since it changes how people live and work in a newly constructed setting.

The four water episodes in this section speak to that sense of things. The Philadelphia waterworks spawned the development of city-water water-supply systems throughout the United States in the early nineteenth century. The Erie Canal was an awe-inspiring water network which revolutionized travel, enhanced trade, promoted urban growth, and provided a way to communicate new ideas. The building of the Toronto Waterfront demonstrated how people have transformed their coastal landscape to better connect to the sea—at least in terms of commerce and trade. And the exploitation of Niagara Falls re-volutionized tourism and energy production—two seemingly incongruous outcomes. In all of these water episodes, the physical environment was transformed, and the respective societies were compelled to adjust to a new era in lifestyle, economic forces, and social challenges.

Community-wide water-supply systems developed slowly in American cities, in 1801 Philadelphia became the first to complete a waterworks and municipal distribution system sophisticated even by European standards. The system was an anomaly, however, since it did not spark an immediate nationwide trend. More than any other sanitary service, an efficient water-supply system was a key factor in the health and well-being of urban populations. As the first city-wide service, it also set the standard for the municipal management of a wide array of other services, such as transportation, communication, health, and policing.

Maintaining a clean water supply for its citizens was an extraordinary task for city leaders. Water going into homes and businesses and out again was no simple task. With piping-in of fresh water came the need to deal with was-tewater which required a whole different system. Equity of service and

DOI: 10.4324/9781003041627-13

protection from pollution were fine ideas on paper, but often were neglected or ineffectively address. A good source of American urban sanitary services is Martin V. Melosi, *The Sanitary City: Environmental Services in Urban America from Colonial Times to the Present* (2008).

Although not part of the Philadelphia case study, the experience of Memphis, Tennessee, in 1870s and 1880, is a good example of the failure in providing equitable sanitary service. Memphis had started from scratch building its sewer system, regarded as an essential weapon against disease. When yellow fever epidemics swept through the city in the 1870s, Memphis only had a few miles of private sewers in the central business district. The poorer classes—mostly black—lived in badly constructed houses in the least clean parts of town. The crisis in Memphis became national news, and city leaders were quick to blame the minority population for the epidemics, citing the unsanitary conditions of their neighborhoods as the cause. Ignorance about how disease was transmitted and the overall lack of attention to city-wide sanitation were the source of the problem. A new sewerage system vastly improved conditions in Memphis by 1880, but recognition of the cause of the city's health problems was not widely understood or acknowledged.

The story of the Erie Canal provides an opportunity to evaluate the nation's biggest infrastructure project in the nineteenth century. In many respects, the objective of "harnessing" rivers to improve on Nature and bend them to the will of humans precedes the decision to build the canal. The fundamental question was: How do we manipulate our water resources to meet our goals—in this case building commerce, shortening travel time, and accessing land and other resources on a larger scale than possible before. An important part of the story is understanding the will to accomplish that task and how the building of the canal could be achieved from a financial and engineering standpoint. Outcomes of completing the canal were based on concrete goals of trade, commerce, and economic success, but short on trying to understand the physical ramifications (and cost of human life) required to complete it. Stories of the pyramids or the Brooklyn Bridge carry similar baggage. In the case of the Erie Canal, the utilization of the resource of water had a singular, finite purpose: to fill the canal with water. Anything else was put aside.

The construction of the Toronto Waterfront is just one example of numerous attempts worldwide to remake the ocean coastline or the water's edge along lakes and rivers. On New York City Harbor in the nineteenth century, water lots and marsh filling added 137 acres of land to Lower Manhattan. The Dongan Charter (1686) set the way for the major alteration of Lower Manhattan. English Provincial Governor Thomas Dongan granted to the corporation of the city "all the waste, vacant, unpatented, and unappropriated lands within the city and island, extending to low-water mark in all parts" which established the limits of the city to land exposed at low tide (Bone, 2004, 156). It also allowed vacant land in Manhattan to be developed by private users. This resulted in the city providing "water lot" grants on parcels laying between marks of high and low tide (essentially land under water half of the time). The grants were used to extend land along the shallow shoreline to provide space for warehouses and as mooring places for larger vessels. This was called "wharfing out" by the English colonists. Soon after the Dongan grant reached its physical limits (1720s), the Montgomerie Charter (John Montgomerie was the colonial governor of New York from 1728–1731) broadened the definition of exploitable underwater land. This allowed for the expansion of Manhattan further out into the bay (Steinberg, 2014, 22–38; Buttenwieser, 30–32, 37, 48–49).

The situation in Toronto was different. The primary goal was not extending the harbor but crafting the port to meet maritime commercial needs. Over time, those needs and wants changed, and competition from the railroads affected the economic life of the city. Nonetheless, terraforming the coastline was done in such a way as to maximize the use of the harbor, sometimes at the expense of the coastal environment.

Niagara Falls has attracted a great deal of attention among historians not only because of its natural beauty but because of the extraordinary ways Americans (and Canadians on their side) have exploited this precious re-source. Tourism became a major, and unpredictable, economic opportunity, but one which made the site crass and artificial. Hydroelectric power de-velopment was much too enticing, and once the technology became of age, the Falls were harnessed much like Lake Eire was harnessed. Along the way, some people saw the value in preserving the esthetic of the place, but success proved uneven.

Taken as a whole these four water episodes provide a glimpse into the emerging urban-industrial society developing in North America in the nineteenth century. Water is a medium serving many purposes once again.

Notes

Bone, Kevin, ed (2004) *The New York Waterfront: Evolution and Building Culture of the Port and Harbor* (New York: The Monticelli Press, rev. ed).

Buttenwieser, Ann L. (1999) *Manhattan Water-Bound: Manhattan's Waterfront from the Seventeenth Century to the Present* (Syracuse, NY: Syracuse University Press, 2nd ed).

Stearns, Peter, and John H. Hinshaw (1996) *The ABC-CLIO World History Companion to the Industrial Revolution* (Santa Barbara, CA: ABC-Clio).

Steinberg, Ted (2014) *Gotham Unbound: An Ecological History of Greater New York, 1609–2012* (New York: Simon & Schuster).

9 Philadelphia's Waterworks: Pioneering Clean Water for Cities

Prior to the 1830s, many American cities faced poor sanitary conditions and suffered crippling effects of epidemic disease. While some of the earliest city-wide water-supply systems appeared at this time, few communities could boast well-developed technologies on the order of those constructed several decades later. Much of the responsibility for urban water and sanitation practices rested with the individual. Philadelphia developed the first city-wide water supply system in the United States in 1801 although it remained somewhat of an anomaly for three decades.

In the mid-nineteenth century, English experiences in particular with sanitation and water-supply systems going back several decades began to influence American cities. Prior to that time, most municipal officials did not detect problems that led them to seek alternatives to existing approaches, such as public wells. Many American towns and cities were on the cusp of change as the nineteenth century unfolded, but only under unique circumstances did rudimentary water-supply systems begin to appear.

The fear of fire and epidemics were great motivators for change. The old "bucket brigade" was grossly inadequate when whole blocks of homes and shops were endangered by fire. Prior to the completion of Philadelphia's system in 1801, it took the bucket brigade fifteen minutes to fill one fire engine with water; after the system was in place it took 1.5 minutes.

The hydrant became the modern symbol for fire protection since it meant that water would be promptly available and abundant to fight a major conflagration. New York, which took a leadership role in fire protection, did not install hydrants before 1830. While hydrants were important in emergencies they also increased the use of water, making a large supply was even more necessary.

Before the turn of the nineteenth century, most cities and towns depended on a combination of water carriers, dipping pails in wells, or draining cisterns to meet their needs. Aside from wells that captured groundwater, large supplies came from local springs, ponds, and rivers, but they were not connected to extensive distribution systems. Even during the first several decades of the nineteenth century, many larger cities and smaller towns continued to rely exclusively on local sources of supply. Unless they hired water peddlers, each citizen used no more than three to five gallons per day.

DOI: 10.4324/9781003041627-14

Fear of fire and epidemics alone was insufficient to lead towns and cities to abandon traditional sources of water and familiar methods of acquiring it. A community needed a political commitment, fiscal resources, and access to new technology. Prior to the 1830s or so, only about half of the major cities and towns had some type of waterworks. The great majority were located in the Northeast, with considerably fewer in the Old Northwest and Upper South.

In 1801, Philadelphia became the first to complete a city-wide waterworks and municipal distribution system sophisticated even by European standards. The Philadelphia waterworks also was an incongruity, since it did not spark an instant nationwide trend. Unique health, economic, and technical factors converged at this particular time to produce what became a model for future systems.

Concern for health, in particular, prompted the campaign for a waterworks in Philadelphia. While the fear of fire always loomed, the startling impact of an epidemic increased public pressure for improved water supplies. At the time a vague notion that "bad" air and water might contribute to sickness led people to believe that the use of the senses—smell, appearance, taste—offered the sole tests of purity. Despite uncertainty in determining disease causation, the seeming correlation between pure water and good health was nevertheless a driving force in dealing with epidemics. *Scott's Geographical Dictionary* described the water in the densest areas of the city as having "become so corrupt by the multitude of sinks and other receptacles of impurity, as to be almost unfit to be drank."

Finding a new source of water arose earlier in Philadelphia, but ravaging yellow fever attacks in 1793 and 1798 led political and business leaders to form a Watering Committee to deal with epidemics. The consensus was that polluted water from wells and cisterns caused the fever and that the city's private wells should be replaced by a community-wide system. The waterworks also could offer needed water to clean the streets, fight fires, and add to the esthetic quality of the city through public fountains.

After examining various options, the committee accepted the proposal of Benjamin Henry Latrobe. The English-born engineer was also a practicing architect, who later became known for his work on the U.S. Capitol (1802–1817). Latrobe recommended a system to pump water from the nearby Schuylkill River and distribute it through mains made of bored logs. He proposed that water would be moved by a steam engine along the river up to a tunnel running under the streets and then by gravity to a pump house at Centre Square in the city. Another steam engine at Centre Square would pump the water to reservoir tanks at the top of the building and then flow by gravity through the distribution system.

Latrobe began the task in 1799 and completed it in 1801. His esthetics as an architect permeated the project. As one historian noted, "The Philadelphia Waterworks at Centre Square was an early example of Latrobe's influential neoclassical architectural style. The building was admired for its proportions and use of Greek prototypes" (Gibson, 1988, 9).

Even after full operation, the machinery never worked as planned. The cost of the system was high, the amount of water pumped was limited, and recurring yellow fever epidemics in 1802, 1803, and 1805 alarmed the citizens. In 1811, the Watering Committee replaced Centre Square with a larger plant in a different location. The plan of engineer Frederick Graff, Latrobe's former assistant, called for a pumping station along the Schuylkill at the foot of Fairmount rise (beyond the city limits), with construction of a reservoir on top of the hill in the city. The new facility was completed in 1815. Steam pumps again were employed but converted to more reliable water power in the 1820s. The Fairmount Waterworks served Philadelphia until 1911.

Figure 9.1 View of Waterworks, Looking South. Fairmount Waterworks, Aquarium Drive, Philadelphia, Philadelphia County, PA. HABS. Public Domain. Wikimedia.

In distributing the water, the new system in Philadelphia first relied on wooden pipes and eventually iron pipes. From the seventeenth century until well into the nineteenth century, wooden mains were commonly used within American cities. The first wood conduits were probably laid in Boston in 1652. Winston-Salem, North Carolina, purportedly built the first citywide system with log pipes in 1776. While rotting and leaking were chronic problems, wooden mains had one advantage—in case of fire, a hose could be connected directly to the main simply by drilling a hole in it. (After the crisis had passed, a wooden plug could be driven into the main. This practice probably was the origin of the term "fire plug").

Not without its flaws, Philadelphia's waterworks was considered by many to be the most advanced engineering project of its time. Ultimately, the city had a system with a much greater capacity than existing demand unlike comparable cities, such as New York, Boston, and Baltimore. To promote its use, clients were initially offered free water for several years. Despite the dread of epidemics, many citizens had not been completely convinced to give up "their cold well water for the tepid Schuylkill water." By 1814, however, 2,850 dwellings were receiving water from the new system.

Philadelphia's example for city-wide systems was widely publicized, but a national trend of adoption did not occur until late in the century. Inexperience in dealing with such a major project, in part at least, helps to explain why urban population growth exceeded construction for so many years. In 1800, there were 17 works for an urban population of 322,000; in 1830, there were 45 works for 1,127,000 urban Americans. "Municipal governments in the nineteenth century were just emerging as effective governing bodies; often the decision to obtain a water system was the first major undertaking of a city government and the first which required a large initial outlay financed by bond issues" (Anderson, 1980, 1).

Rural-dominated state legislatures often attempted to check city growth by controlling services from the state capital or restricting the taxing and financing power of the city in its charter. Thus, it was exceedingly difficult for cities to provide services, even when they accepted responsibility for them. "Home rule" for many cities did not become a reality until late in the century. Not surprisingly, almost every city and town initially turned to private agents or companies to supply water.

Private companies received franchises through corporate charters, which was a typical way to generate public-works activities in the eighteenth and early nineteenth centuries. Since few companies could meet the expectations of the cities for good service, plentiful and pure water, and low price, those who agreed to do so received franchises with substantial concessions. It was not unusual for a franchisee to get a long-term contract, exclusive rights to supply water, the right to acquire property by eminent domain, exemption from taxation, and other benefits. In 1800, 16 of 17 American waterworks were private, and 36 of 45 in 1830.

In New York City, a freshwater pond had been the major source of supply even after 1800, while water from many private wells suffered saltwater infiltration and pollution from privy vaults, cesspools, and street drainage as early as 1750. In 1774, the Common Council contracted with an English engineer to build a municipal system using a steam engine to lift water into a central reservoir. The Revolutionary War derailed the project, and not until 1799—after a devastating yellow fever epidemic—was it renewed. City leaders realized that rivals Boston, Philadelphia, and Baltimore were building or were proposing to build waterworks of their own.

Water quickly became the focus of a major political struggle in New York. The council requested the legislature to provide it with special powers to establish a water system. Assemblyman Aaron Burr maneuvered to acquire a charter for a new private water company—the Manhattan Company—instead of supporting the development of a municipal system. The perpetual charter granted the company wide powers with few obligations, and Burr was intent on using it to amass surplus capital in the hopes of building a great banking business.

From the vantage point of water-supply service, the company was modestly successful. At its peak, it provided water for only one-third of the city and was continually embroiled in controversy. Between 1801 and 1808, Burr faced staggering political setbacks, including losing control of the Manhattan Company. He was dropped from the board in 1802, making way for the rise of his political rival, DeWitt Clinton (of Erie Canal fame). Clinton soon realized that meeting the demand for city-wide distribution of water was impossible within the existing system. Discussions about selling the company to the city became more frequent, but in the short term the charter was revised and the company continued its favored position.

The deteriorating quality of the water supply weakened the Manhattan Company's hold on the city's water service. In 1825, a bill granting a charter to the New York Water-Works Company was enacted. Controversy over its charter, the lack of good supplies of pure water in the immediate area, and the pressure from the Manhattan Company and other rivals ended the short-lived venture.

The basic requirements for good, accessible sources of water were not met until the completion of the Croton Aqueduct and Reservoir in 1842, which provided water from upstate New York for the first workable municipal system in New York City. The project is regarded as a sublime engineering feat, and as a symbol of the conquest of nature in service to the urban population explosion. The Croton Aqueduct project is also an important example of changes in the scale and complexity of modern water-supply systems. In a rare moment of political harmony, the voters, the state legislature, and the New York Common Council agreed to construct the forty-one-mile aqueduct from the Croton River in Westchester County to New York City. Tapping distant water supplies was not without its tensions, especially between the cities attempting to acquire the water and the locals wary of the exploitation. A notorious example occurred in the early 1900s, when Los Angeles successfully redirected water from the Owens Valley, leaving valley residents without sufficient water for their crops.

Boston also endured a long water-supply debate before it developed an adequate system in the 1840s. From 1630 to 1796, the city derived all of its water from wells and cisterns, and the quality was dismal "hard, highly colored, often odorous, saline, bad-tasting, and sometimes polluted" (Nesson, 1983, 1). In 1796, Gov. Samuel Adams approved an act creating the Aqueduct Corporation, which built a line from Jamaica Pond in Roxbury to the city. The distribution network was extended in 1803, but it did not provide service for the entire community. There was no further attempt to improve the existing system until 1825, but civic leaders chronically argued over the water supply through the mid-1840s.

The pattern in the Midwest, the South, and elsewhere was similar to experiences in the Northeast. Some of the larger cities made the transition to city-wide systems early, with most cities and towns following more slowly. Cincinnati was the first "western" city with a waterworks. In 1813, community leaders contracted to drill "possibly 30" public wells in a single season. However, an 1817 ordinance chartered the Cincinnati Manufacturing Company to develop a system, one of the earliest such concessions granted. In 1839, the works were sold to the city. At the time of the purchase, the property only consisted of a pumping station and reservoir grounds. The company had run into financial trouble in the intervening years, and chronic problems in meeting their obligations stimulated distrust of the company and kept alive the possibility of public control.

The St. Louis Water Works was built in 1830. In 1821, a general concern about fire hazards led to a demand for a better water supply. Finally, in 1829, the city council offered a $500 prize for the best plan. Within a short period of time, the city signed a contract with Wilson and Company and the work on the installation began in 1830, but the water did not move through the pipes until the 1840s.

Benjamin Latrobe brought his innovations from Philadelphia to New Orleans, which was on the brink of a major growth spurt. His plan was to secure a franchise for himself and his investors to turn profits from the sale of water. The New Orleans waterworks was similar to Philadelphia's in several respects. A steam engine would pump water from the Mississippi River through a pipe into six elevated wood reservoirs. Gravity would carry the water through a combination of wooden and iron pipes. Benjamin's son Henry completed drawings of a fountain which was never built along the riverfront square and fire hydrants and hand pumps at various intersections.

Benjamin Latrobe was known for accepting projects that he was not immediately prepared to undertake. And since he was occupied in the Northeast, he sent Henry to the Crescent City in 1811 to begin the work. He himself did not arrive on the scene until 1819. Aside from the English engineer's absence, technical setbacks, and problems with investors, the Latrobes also had to contend with the disruption of the War of 1812. To his credit, Henry kept the project from unraveling. But in a major blow to his father and to the project, young Henry succumbed to yellow fever in 1817. Until Benjamin arrived in New Orleans, one of Henry's associates took charge. Upon completion of most of the work on the waterworks, Benjamin himself apparently contracted yellow fever and died. Within a year, New Orleans Water Works Company, struggling to survive, was sold to the city. This project was Benjamin Latrobe's final engineering legacy.

The technical achievements in developing early water-supply systems had some bearing on levels of consumption in the early nineteenth century. The new systems, however, did not provide for equity of service. In the mid-1820s,

Cincinnati had more than 26,000 feet of wooden pipes but served only 254 industrial and home users. At the time, the daily consumption of water probably averaged between three and five gallons per capita, with higher consumption by those who could afford to purchase additional supplies.

For all the improvements begun by private companies through franchises, accessibility to the water supply was still largely linked to class. Affluent neighborhoods and the central business district received the lion's share of water, while the working-class and poorer districts (including minority communities) often relied on polluted wells and other potentially unhealthy local sources. While Philadelphia was the pioneer in waterworks, it took a generation and longer for owners of slum property to install plumbing for the poor.

The crucialness of adequate supplies of water to meet the needs of citizens, commercial establishments, and industry—and the mandate of cities to protect public health—meant that authorities in the largest urban areas wanted centralized systems under their direct control. Boosterism was an additional motivator since an effective water system was a powerful promotional tool to enhance a city's economic base. While many water companies had been profitable, capital investment in the more modern systems was steep, and operating costs were on the rise. Private service, therefore, was gradually phased out in several communities. In addition, public control of the water supply enhanced the authority of city government vis-a-vis the legislature or rival cities, thus private owners often were under pressure to sell out.

The desire of city leaders to convert private systems into public or to build new public systems, rested on more than the will to do so. The central issue was the ability of cities to incur debt to fund major projects and to sustain the high costs of operating these new technologies. As the nineteenth century unfolded, city finances underwent changes in scope and complexity which ultimately made the development of public systems achievable. The power of taxation was central to a municipality's budgetary process. In cities throughout the United States, ownership rather than income became the basis for taxation.

The general property tax emerged as the basic mechanism for raising revenue. Unlike England, where local taxation was based on the rental value of property, in the United States, total property wealth became the standard. While property tax was the most important source of local income by mid-century, taxing powers of the cities usually were controlled by the states. In some cases, the state imposed added responsibilities on city government with no additional financial support. For example, epidemics of the period led cities to request state aid, but without much success.

Especially in the case of capital improvements, like water-supply systems and sewerage, special assessments were an important financial tool to augment property taxes. As the range of city services expanded, demand for more revenue increased. The requirements of urban growth meant that reliance on taxes and special assessments was likely to satisfy neither the citizenry nor the local business establishment. Increased municipal debt became much more common in the nineteenth century. By 1870, the popular referendum was a tool for obtaining public support for bond issues. Most significantly, methods of taxation and the allocation of municipal funds prevalent by 1870 set the pattern for urban finance for more than a century.

In developing citywide water-supply systems, substantial public investment proved difficult for all but the largest and most fiscally sound cities. If the legislature was not withholding extension of greater authority, the council was debating the wisdom of increasing the city's bonded indebtedness or engaged in partisan debate. Going back at least to 1855, the percentage of public water systems tended to vary with the general financial health of the cities, at least until the 1880s when other issues also influenced the decisions. In addition, municipal indebtedness had steadily grown in order to finance several improvements, including water supply. By 1860, municipal debt was three times the federal debt and almost equal to the aggregate state debt.

Liberalization of charters and other fiscal changes provided opportunity for cities to finance water-supply systems and other public works, especially beginning in the 1860s. In most cases, a combination of local circumstances and the experience of other cities influenced the shift from private to public. Outside of a core of emerging major cities in the industrial East and a few others sprinkled throughout the country, the transition from private companies to municipal service more typically occurred in the late 1860s and after.

In the South, municipal water systems were rare in this period. In Reconstruction-era Atlanta, plans moved ahead for a new waterworks, but its primary thrust was for fire protection and to serve business and industrial needs. Without a creditable municipal water supply, the more affluent turned to purchasing spring water from a vendor or depended on personal wells. In black neighborhoods drainage was poor, sewer outfalls often dumped wastes there, and wells were badly polluted. Likewise in Memphis, little attention was given to residential water service.

The American cities that turned to community-wide approaches set patterns for more elaborate sanitary services in the future. The application of the initial technologies ran ahead of an effective understanding of the causes of disease. This became clear after the turn of the twentieth century with the acceptance of the idea of disease transmission through bacteria (germ theory) rather than because of bad smells or miasmas (filthy theory). In time, advances in chemistry helped clarify how some diseases could be contracted through pollution. Eventually, the building of newer systems demonstrated how attention to revised health theory could improve the quality of water delivered to the citizenry.

By 1880, modern citywide systems provided greater quantities of clean water over sizable areas and incorporated rudimentary safeguards to insure purity. A growing preoccupation with water quality—a direct result of the sanitary movement of the nineteenth century—was bringing attention to filtration techniques and new methods of water treatment. City leaders and sanitarians alike were demanding more from their water-supply service than convenience at the tap.

Various approaches to assure a pure water supply and a reduction in waterborne disease emerged from bacteriological and chemical laboratories. But the first means of water purification readily available to cities in the late nineteenth century—one that fit within the framework of the filth theory—was filtration through sand or gravel to improve water clarity, odor, and color. Albert Stein, the designer of Richmond's waterworks, was the first to attempt to filter a public water supply in the United States in 1832. Pumping water from the James River, Stein prepared a sand filter in the reservoir, but he could not get it to operate effectively. During the next forty years, several major cities, including Boston, Cincinnati, and Philadelphia, considered installing sand filters, but they were too expensive at the time.

A major step forward, but not recognized immediately, was the "Report on the Filtration of River Water, for the Supply in Europe, made to the Board of Water Commissioners of the City of St. Louis" (1869) by Brooklyn engineer James P. Kirkwood. In 1865 Kirkwood had recommended that St. Louis and Cincinnati employ filters in their water systems. However, nothing came of it at the time. Later, he was hired by the City of St. Louis to survey locations for supply works along the Mississippi River. Upon recommendation of a plan that included filtration, the water commissioners instructed him to travel to Europe to get first-hand knowledge of the technology. While away, opposition to Kirkwood's plan mounted which led to a clean sweep of the commission and replacement with members unwilling to underwrite the cost of filtration. The city would not publish his report and did not filter its water until fifty years later.

Kirkwood's report ultimately became a bible for those cities interested in copying the European experiments. For several years following its completion, little additional first-hand knowledge was gathered about the various European systems. By the early 1870s, a few cities began to recognize the value of filtering water. Poughkeepsie, New York, built the first American slow sand filter in 1870–1872 based on the designs in Kirkwood's report. By 1880, however, there were only three slow sand filters in the United States and none in Canada.

Few of its major proponents claimed that filtration was the only answer to disease prevention, but contemporary statistics indicated that it was especially effective in controlling typhoid fever. The results often were so dramatic that people lost sight of the idea that filtration's primary purpose was reducing turbidity and removing suspended matter.

Along with experiments in filtration, a variety of pumping techniques and changes in pipe technology helped to transform older systems into modern, centralized waterworks. Aside from gravity systems, steam pumps were increasingly employed at the source and as a way of moving water to reservoirs, tanks, and standpipes. Wooden pipe was adequate in low-pressure gravity systems but could not withstand the action of high-pressure pumping engines. By 1850, iron pipe came into wider use in the United States, especially in high-pressure systems. In the West, wood was used for large aqueducts, irrigation, hydroelectric plants, and hydraulic mining.

In the early twentieth century, chemical water treatment made important strides. Chlorination became well established; experimentation with copper sulfate to control algae was underway. Public health leaders promoted the idea that typhoid fever and related diseases could be prevented through the use of a combination of filtration and treatment. (See Chapter 18 for more details on chlorination.)

With Philadelphia's lead, the construction of water-supply systems in American cities in the early- to mid-nineteenth century was a major step in providing clean water for many essential purposes ranging from improved health to commercial uses. Soon water was being metered as a way of commodifying the new water service and also controlling usage and limiting waste. As the first city service, water supply's essential role in the life of individuals and their cities was recognized emphatically.

Notes and Further Reading

Anderson, Letty Donaldson (1980) "The Diffusion of Technology in the Nineteenth-Century American City: Municipal Water Supply Investments" (Ph.D. dissertation, Northwestern University).

Gibson, Jane Mork (1988) "The Fairmount Waterworks," *Bulletin of the Philadelphia Museum of Art* 84 (Summer).

Melosi, Martin V. (2000) *The Sanitary City: Urban Infrastructure in America from Colonial Times to the Present* (Baltimore, MD: Johns Hopkins University Press).

Nesson, Fern L. (1983) *Great Waters: A History of Boston's Water Supply* (Boston: Brandeis University Press).

Rawson, Michael (2010) *Eden of the Charles: The Making of Boston* (Cambridge, MA: Harvard University Press).

Smith, Carl (2011) *City Water, City Life: Water and the Infrastructure of Ideas in Urbanizing Philadelphia, Boston, ad Chicago* (Chicago, IL: University of Chicago Press).

Soll, David (2018) *Empire of Water: An Environmental and Political History of the New York City Water Supply* (Ithaca, NY: Cornell University Press).

10 Water Rerouted: The Erie Canal

Before the arrival of the railroad, the only practical means of transportation on land was by foot, horse, and wagon, or on water by human- or wind-powered boat. For many years, water proved to be the most important way to move heavy objects from one place to another. The medium itself was essential. As we have seen, the Aztecs used canals to transport people and goods within Tenochtitlan and outside the city, but on a relatively small scale. Without many—if any—navigable waterways in the old Southwest, indigenous peoples and later Spanish/Mexican settlers had to rely on land travel. The building of the Erie Canal was a revelation in transportation, but it also was an infrastructure that changed the water that ran through it and the social and physical landscape around it.

Human action in designing and constructing canals for inland transport changed travel in immeasurable ways. The first canals in the United States were based on those in Europe, especially in France, England, and Holland. The American canals built in the 1790s were small in size, took a long time to construct, and were financed by private companies limited in funding. A few years later several short canals were dug, including the Schuylkill and Susquehanna Canal in Pennsylvania and the Potomac Canal in Virginia. The "Canal Era" in North America began in earnest with the Erie Canal.

The completion of the Erie Canal in 1825 not only allowed for heavy-duty traffic along a water route from the Great Lakes to the Atlantic Ocean but transformed the area itself, offered an important new path for immigration, and boosted the economy of New York and the United States in general. The canal brought new people to upstate New York seeking economic opportunity and inspired by spiritual renewal and other new ideas. In the early nineteenth century, the canal had helped to spread news of movements such as the evangelical Second Great Awakening. According to one historian of religion, at this time "revivals were so habitual and powerful in the area west of the Catskill and Adirondack Mountains that historians have labeled this ecclesiastical storm center the 'Burned-over District'" (Backmann, 1969, 301).

It is no wonder that the Erie Canal was regarded as "the Great Water Highway," an engineering marvel, and one of the wonders of the world.

DOI: 10.4324/9781003041627-15

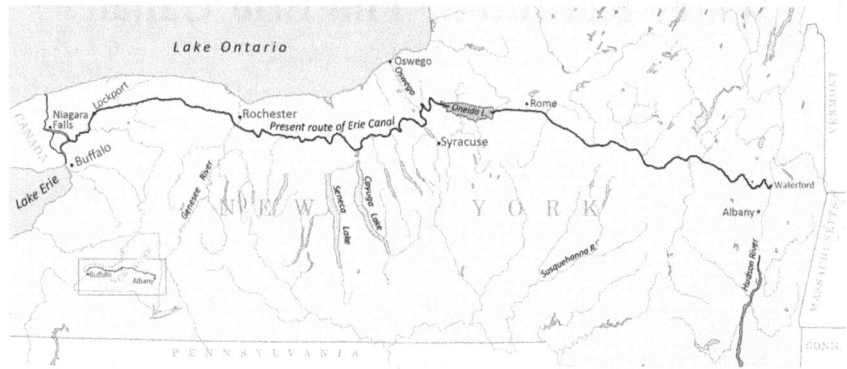

Figure 10.1 Erie Canal Map.

Sources: USGS National Map 1 Million scale data (http://nationalmap.gov/small_scale/atlasftp.html?
openChapters=#chpbound) & the NYS GIS Clearinghouse (https://gis.ny.gov/gisdata/). Rosemary
Wardley. Creative Commons Attribution-Share Alike 4.0. Wikimedia.

For many decades before work on the Erie Canal began, people dreamed
of a direct water route connecting the Great Lakes to New York City (via
the Hudson River) and the Atlantic Ocean and opening the sparsely populated northern sections in the Midwest. Early explorers searched in vain
for a water route from the population centers of the East Coast to lands rich
in resources along the Great Lakes and into the Midwest. The Northwest
Territory (Ohio, Michigan, Indiana, Illinois, and Wisconsin), formed after
the American Revolution, held timber, fur, a variety of minerals, and fertile
land for farming.

Throughout the eighteenth and early nineteenth centuries, it required
weeks of travel by teams of oxen or horses to transport bulky items overland.
The journey from Buffalo to Albany took an eight-horse wagon from 15 to
45 days. The State of New York had tried at first to meet its transportation
problems in the late eighteenth century by chartering private companies to
build turnpikes that would connect various towns along the Hudson and
Mohawk Rivers with inland agricultural areas. By 1821, 4,000 miles of
turnpike had been built. Even with this burst of activity, the turnpikes were
insufficient to carry large volumes of goods cheaply, quickly, and over long
distances.

Until the Erie Canal was constructed, the City of New Orleans dominated
interior trade in North America because of its location on the Mississippi
River. As early as the seventeenth century, the French, Dutch, and British
competed for the rich fur trade on the continent. People began to realize that
New York City (or New Amsterdam before that) with its deep harbor and
access northward along the Hudson could be a competitor for New Orleans
and an ideal site for trade with Europe.

The early basis for migration and settlement patterns and trade routes in New York were influenced by geography and by the experiences of the Iroquois Federation or Six Nations. Most people in New York State over the years lived in an area from New York Harbor north to Albany, and then west towards Syracuse and on to Buffalo at Lake Erie. The various rivers and lakes along the route lie away from the Northeast Appalachian range of the Hudson Highlands, the Taconics, the Catskills, and the Adirondacks. The Appalachian Mountains were the greatest impediment to east/west travel in New York and thus frustrated a variety of attempts to dig a canal from Lake Erie that would eventually lead to the Atlantic.

Attempts to build canals elsewhere on the Eastern seaboard were not very successful, because the elevations were too high. New York had some advantages in elevation and terrain. The area possessed several natural waterways and substantial lowland for passage from the Atlantic Ocean to the Great Lakes. A canal route was possible without scaling or cutting through mountains by going north through the Hudson Valley and west through the Mohawk Valley. About fifty miles north of New York Harbor the Hudson River cut through the Highlands along the Appalachian's 1,500-mile barrier. Approximately 100 miles north of that point, the Mohawk River joined the Hudson, extending 100 miles west to the Oneida Carrying Place across which lies the headwater of Wood Creek. The small stream flowed to Oneida Lake and then to the Oneida River, which joined the Oswego and Seneca Rivers. The Oswego then flowed north to Lake Ontario. Most of Lake Erie lies to the southwest of Lake Ontario, with Buffalo at its most eastern point.

The two natural breaks in the Appalachians were the obvious points to connect the Great Lakes with the Atlantic. The first was a gap where the Hudson River flowed from its source at Lake Tear of the Clouds south through the Highlands, thus providing a sea-level access between the Great Valley of the Appalachians and the sea. The Hudson ran at sea level from its mouth near New York Harbor to its confluence with the Mohawk River (which allowed for ocean-going vessels to reach the state capital in Albany). The trick was to connect the lower Hudson with the natural break near the Highlands and then move north.

The second break was a wide opening between the Catskills and the Adirondacks. A major obstacle was Cohoes Falls above the Mohawk River's mouth. From there the Schenectady River was not navigable. Beginning in the late eighteenth century, efforts to improve New York's water route through either break resulted in limited success or several false starts.

In 1724, Cadwallader Colden, New York surveyor general (and later colonial Governor of the Province of New York), wrote a report for the governor describing transportation possibilities between Albany and Montreal, Canada, and between Albany and Lake Ontario. Not until another hundred years had passed was the Corresponding Association for the Promotion of Internal Improvements organized to promote the building of a canal across

New York. But in the meantime, rapid canal building in Great Britain during the 1760s and continuing interest for canals in New York kept the idea alive.

In 1785, an Irish immigrant and engineer, Christopher Colles, called for the state legislature to create better navigation on the Mohawk. In 1792, the Western Inland Lock Navigation Company began clearing the riverbed and built a small canal and lock system to bypass the river's falls and rapids. While not a complete answer to improving water transport across the state, the work allowed for boats to travel between Oneida Lake and Schenectady. Portage (carrying by land) still was required to Albany and only roads were available west of Oneida Lake. The semi-private company never completed all its plans, constantly facing funding issues, charges of corruption, and partisan politics. In 1820 the state bought the company's works and ended its activities.

Jesse Hawley, a flour merchant, made one of the most influential arguments for constructing an east-west canal across New York. Hawley had gone bankrupt because he could not find an effective way to get his product to market. While in debtor's prison in Ontario County he published a series of fourteen essays in the *Genesee Messenger* in 1807–1808 under the name "Hercules." Hawley offered practical information on the route, cost, and the benefits of "connecting the waters of Lake Erie and those of the Mohawk and Hudson rivers by means of a canal."

The timing of Hawley's remarks was significant, just when politicians and business leaders alike were discussing the need for internal improvements (public works) to help boost the economy after the Revolutionary War. President Thomas Jefferson opposed federal involvement in developing a canal, having just signed a bill to construct the National Road to connect Baltimore to St. Louis. He believed that building a long canal in what constituted wilderness was "a little short of madness to think about it."

As a result, the State of New York became the focal point for action, but not without a good deal of political in-fighting. In 1808, the State Legislature agreed to hire a surveyor, James Geddes, to explore an interior route for the canal. In 1810, the State Senate established a Canal Commission to propose a canal linking the Great Lakes and the Hudson River. Two routes were considered, a Lake Erie route and a Lake Ontario route. The former was viewed as preferable because many believed that traffic entering Lake Ontario would continue down the St. Lawrence River to Montreal, and away from American communities. But federal funds to build the canal from Lake Erie were not forthcoming, even though the state had received large donations of land to help finance it. Too many interests were at loggerheads to come to a consensus.

The War of 1812 delayed further discussion about a canal, and by 1815 many believed the idea had died. This conclusion was reinforced in 1816 when President James Monroe vetoed the "Bonus Bill" that would have provided federal money for the Erie Canal. Madison stated that he was not convinced that federal support for the canal was constitutional without a specific amendment authorizing Congress to fund transportation projects.

Some saw his motives as sectional bias (given that he was a Virginian), but the offshoot of the decision was that states, temporarily at least, retained power over such decisions. Nonetheless, the war had stirred strong nationalist feelings and several leaders argued that strategic interests of the United States made clear that an inland transportation system was necessary for the new nation. This was particularly important because the British were entrenched in Canada.

State money for the canal was substituted for federal funding largely through the efforts of DeWitt Clinton, who campaigned for it during his run for governor in 1817. While New York City mayor, Clinton had been persuaded by the Hawley essays that the canal was imperative. Aside from political battles with Federalists, Clinton's Democratic Republicans split over the canal issue in an intense fight. Some landowners in southern New York opposed being taxed for a project of benefit to other parts of the state (other regions also seemed threatened). Clinton's opponents, drawing support from these landowners, eventually became the nucleus for Tammany Hall, which ruled New York City politics for much of the nineteenth century.

Under Governor Clinton, a bill passed the legislature in April 1817 that provided for the construction of two canals: the Western Canal (or Great Western Canal, later the Erie Canal) connecting Lake Erie to the Hudson, and the Northern Canal to connect Lake Champlain with the Hudson. Construction of the Erie Canal began on July 4, 1817, on the Mohawk River at Rome. Opponents referred to the project as "Clinton's Big Ditch" or "Clinton's Folly," but it proved to be an historic achievement. The canal was built in three parts and designed to run from Buffalo to the Hudson River at Albany, a waterway of 363 miles with 83 locks. The canal did not utilize any natural lakes or rivers in its design; thus, it was entirely contained. Once completed, boats on the canal would be pulled along by draft animals on tow lines along a ten-foot-wide path. Steam-powered boats eventually replaced the original method.

The most effective way to build the locks was to use stone and cement. The mortar from cement would harden and create a watertight surface. Ideally, limestone rock from Europe was best suited for the task, but it would have to be imported at great cost. A compromise solution was to use as much of the local resources as possible for the locks with the outer edges of the mortar joints made with imported hydraulic cement.

By 1819, the middle section of the canal (extending 98 miles between Utica and the Montezuma marshes strewn with quicksand) was completed. The next year, transportation on that section was underway, most of the western section was completed, and construction of the eastern section had begun. By 1822, the canal was open between Schenectady and Rochester, and the next year the junction with the Hudson was made. In 1823, the Champlain Canal was finished, which ended Montreal's control of northern New York's economy.

On October 25, 1825, the entire length of the canal was completed—two years ahead of schedule. Every stage of construction was celebrated.

On October 26, canon fire could be heard along the whole distance of the canal from Buffalo to New York City as a flotilla of boats moved east from Buffalo led by the *Seneca Chief*, which carried Clinton and other officials. At the end of the journey, a massive fleet congregated in New York Harbor. Clinton took a cask of water from Lake Eire and poured it into the Atlantic at Sandy Hook calling it "The Marriage of the Waters."

The work to build the canal had been arduous. The locks would manage a rise in elevation of 500 feet from the Hudson to Buffalo. The canal itself was 40 feet wide and 4 feet deep along its 363 miles. Overhead, eighteen aqueducts were used to let streams cross the canal. Small contractors handled most of the actual construction except for the ascent through the ridge which required blasting through solid rock. The multiple contractors—many were wealthy farmers—dug small sections of the canal averaging about three miles, supplying their own equipment and labor. Massive reservoirs were built to ensure that the canal had a constant supply of water.

The engineers who designed the canal were amateurs (surveyors, lawyers, teachers), learning as they went, and came to be nicknamed the "Erie School of Engineers." Benjamin Wright (a judge and a surveyor) was chief engineer, and James Geddes (lawyer and surveyor) was assistant chief. John B. Jervis, who had chopped wood for a living and served as a surveyor's aide, ultimately worked on the Delaware and Hudson Canal, the Croton Aqueduct, and some early railroads. Canvass White discovered cement that grew harder underwater and applied the product to wetland construction.

At the beginning of the project, there were no professional civilian engineers in the United States. And there were no engineering schools in the country—save the growing importance of the military academy at West Point and the U.S. Army Corps of Engineers. But the project did spur the development of several new programs that provided needed expertise for future internal improvements. Rensselaer Polytechnic Institute was established in Troy and trained many engineers during the latter stages of the canal-building era.

Approximately 3,000 individuals worked on the canal at any one time. Many of the laborers were newly-arrived immigrants from Ireland, Wales, and Germany as well as locals living near the canal, who did much of the back-breaking work with picks and shovels and, in some cases, horses. Workers were paid by the cubic yard of dirt removed and had to cover their own expenses. Dirt was hauled away on small, awkward carts. Before a horse-drawn stump puller, plow, and scraper were introduced, forest areas required exceptional hard work to clear. Steam machinery was not yet available.

Camp life was harsh for the workers. In the areas with marshy bogs, they faced infestations of mosquitoes in the summer and suffered a variety of maladies over the whole period of construction. There was significant illness and loss of life due to malaria, smallpox, cholera, and yellow fever. Potential injuries from landslides and explosions were always present. As historian Edward Countryman noted, "The Aztecs of pre-Columbian Mexico had accomplished engineering feats on such a scale, but nothing like the Erie Canal's works had been achieved by anyone in the British colonies or in the young United States" (Klein, 2001, 278).

The impacts of the canal on the state and national economies, on immigration, and on urban development were stunning. Because of the canal, carrying of goods changed the direction for western commerce from Southern ports (via the Ohio and Mississippi Rivers to New Orleans) to the burgeoning industrial North. This helped shift western political allegiances away from the slave-holding South to support the Union in the Civil War.

Goods from the west could cross the Great Lakes to Buffalo, then be loaded on boats on the canal to Albany and New York City, and feasibly then on to Europe. Canal boats could carry 30 tons of produce, which lowered the cost of transporting them from Buffalo to New York City from $100 per ton to less than $10 per ton. Freight speed also was substantially reduced to about 55 miles per twenty-four hours. (Express passenger service moved at about 100 miles per twenty-four hours; a trip from New York City to Buffalo via the canal now would take about four days rather than weeks). Traffic from New York City to upstate New York State also proved fortuitous. Such products as fresh oysters—not available upstate at the time—could be shipped from the Atlantic to Albany, Rochester, Buffalo, and elsewhere. By 1853, the Erie Canal carried sixty-two percent of all trade in the United States.

In just nine years, toll charges on the canal were enough to repay the state loan and helped finance several branch canals and enlargement of the Eire Canal. In 1824, traffic on the finished sections of the canal produced about $300,000 in tolls, and some predicted that in two years revenues would reach $1 million—although it fell short of that. Between 1826 and 1835 annual revenues doubled, and then more than doubled in the next twelve years. In 1847, peak revenue was over $3.3 million. Because of the success of the project, in 1882 the toll charges for the Erie Canal were eliminated.

Although increase in tolls was impressive, the great rise in tonnage on the canal was even more significant. In 1825, 218,000 tons of goods were transported on the canal, reaching 4,650,000 in 1860. The canal had its financial ups and downs, especially after periods of excessive building and economic shocks such as the Panic of 1837. The post-Civil War years saw the culmination of a period of economic decline along the canals, especially with competition from the railroads beginning in the 1840s.

Travel westward to acquire or provide goods and products was made much easier as was a path for settlement on the frontier. Extensive growth of economies like large-scale grain farming rapidly emerged. Several cities grew on the shores of the Great Lakes: Cleveland and Toledo along the south shore of Lake Erie; Detroit and Windsor, Ontario on the western end; Bay City, Michigan on Lake Huron; Green Bay, Milwaukee, and Chicago on Lake Michigan; and Thunder Bay, Ontario and Duluth on Lake Superior.

Commercialization of agriculture was possibly the most immediate consequence of the canal-building movement. The cash market (farming for profit) was underway before the Erie Canal but grew mightily. A barter economy for farmers still existed especially in the northern parts of New York and beyond the Appalachians, but it was declining.

Falling freight rates (first with canals and then with railroads) boosted commercialization. In some areas, it also led to rural decline and migration of farmers to the cities. Centralization of farming in the hands of the few sometimes led smaller farmers to take on cattle and sheep grazing and dairy production. On the largest scale, the changes in the state's economy (and the other areas touched by the canal and eventually the railroad) created greater economic diversity including lumbering and quarrying, and attracted more outside investment. The transporting of iron ore and coal throughout the Great Lakes system accelerated industrial development. Indeed, the canal helped increase access to coal reserves in Pennsylvania, thus reducing dependence on foreign sources.

Figure 10.2 Original lithograph showing the Erie Canal at Lockport, New York c. 1855. Published for Herrman J. Meyer, 164 William Street, New York City. Public domain. Wikipedia.

Several small towns grew into substantial cities along the canal route. Rochester became an important center for boat building and the "Flour Capital of the World;" Schenectady for electrical equipment; Utica for copper wire; and Syracuse for salt. Buffalo became a major transshipment point for farm produce from Ohio, Indiana, Illinois, Michigan, and Canada. By 1835 it was the largest inland port in the world. In recent years, about 80 percent of upstate New York's population lived within 25 miles of the canal route.

New York City itself saw a giant leap in commerce and industry, enhancing its development as a major international trade and financial center, as well as "the Granary of the World." It became the focus of international arrivals, ultimately reaching more than 85 percent of the total. Between 1820 and 1850 its population quadrupled.

Almost as soon as it was completed the Erie Canal proved too small. In 1835, the state decided to enlarge the canal to 70 feet wide and seven feet deep (and to reduce the number of locks to 72). When completed in 1862, boats now could carry up to 100 tons of produce. However, the canal did have to close to shipment for up to five months each winter. In 1903, the state again sought to increase the canal's dimensions to about 120–200 feet wide by 12–14 feet deep.

By 1918, the entire system was completed and was 338 miles in length from Waterford to Tonawanda. It was renamed the New York State Barge Canal (later the New York State Canal System) and connected the main channel of the old Erie Canal with Lake Champlain, Lake Ontario, and the Finger Lakes. The new waterway could handle large steamships. By this time, railroads and the St. Lawrence Seaway cut deeply into the canal traffic, and most recently the New York State Canalway Trail uses more than 200 miles of the Barge Canal for multipurpose recreation, such as pleasure boating, biking, and hiking.

The New York government also promoted canal development in other parts of the state. The northern portion of New York benefitted from the Champlain Canal (linking Lake Champlain with the Hudson). In 1825, it authorized surveys for as many as seventeen additional canals, with nine projects funded and built. By 1841, there were more than 600 miles of canal being operated with 300 more under construction.

New York also encouraged the development of a competing transportation system—railroads. At first, the railroads were a supplement to the canals, but that rapidly changed. For a short period, the state government tried to protect the financial interests of the canals by prohibiting railroads from hauling cargo in competition with them. (This regulation soon died in 1851). The legislature incorporated the first railroad company in 1826. In 1831, the Mohawk and Hudson line connected Albany and Schenectady. In 1853, the New York Central Railroad was created by joining several smaller railroads. By the mid-nineteenth century, thirty railroads were in operation in New York on 1,649 miles of track, with 1,000 more under construction.

The combination of the canals and railroads created a revolution in transportation in the state and well beyond. Commercial farming and industrial development helped produce the modern market economy.

But the Erie Canal also had an impact on the country that could not be measured only in dollars and cents. In an essay entitled "Governor DeWitt Clinton's Dream" (1825) Clinton declared,

> As a bond of union between the Atlantic and Western states, it may prevent the dismemberment of the American Empire. As an organ of

communication between the Hudson, the Mississippi, the St. Lawrence, the Great Lakes of the north and west and their tributary rivers, it will create the greatest inland trade ever witnessed. The most fertile and extensive regions of America will vail themselves of its facilities for a market. All their surplus productions, whether of the soil, the forest, the mines, or the water, their fabrics of art and their supplies of foreign commodities, will concentrate in the city of New York, for transportation abroad or consumption at home. Agriculture, manufactures, commerce, trade, navigation, and the arts will receive a correspondent encouragement ...

While overly dramatic, several observers have remarked on the way the canal played a role in drumming up nationalist sentiment and becoming a symbol of progress. Historian Ronald E. Shaw remarked, "The Erie Canal was built by the state of New York alone, but it was by its very nature a national enterprise which related the growth of the other states to New York" (Shaw, 1966, 397).

What often gets lost about infrastructural changes like the Erie Canal—particularly one that deals so explicitly with water which is central to this book—is its environmental impact. The canal deserved many of the superlatives said about it (although there were several incidents of political corruption and mismanagement along the way), but the building of the Erie Canal came with a price largely unrecognized in the early nineteenth century. It is an old story of tradeoffs—economic gain for remaking Nature.

For many at the time, the Erie Canal was extending or improving what remained unfinished in Nature. The ambitious project fundamentally transformed upstate New York created an "artificial river," and engineered a new environment (Stradling, 2010, 47, 50). Historian Carol Sheriff stated it well: "For all its technical brilliance, the Erie Canal had created a landscape that people took for granted" (Sheriff, 1996, 172).

Changes took some obvious and some not-so-obvious forms. Thousands of acres of relatively virgin land were converted for agricultural use. Deforestation with accompanying runoff occurred in some locations along the canal route as did waterlogging of some property and diverting of feeder streams. Culvert washouts could disrupt traffic. Draining swampland, at the time viewed as recovering usable land, destroyed animal and bird habitats and degraded various ecosystems. Invasive species moved in and out of canal waters to the Great Lakes and to the Hudson and Mohawk Rivers. Over the years, the network of navigational waterways in the northern part of the state introduced into the Hudson Basin an estimated 32 exotic species including sea lamprey, alewife, and white perch. In some cases, diseases and illnesses could spread more easily.

Simple habits such as dumping waste into the canal fouled water quality, and as cities grew untreated sewage poured into all the nearby waterways. The advent of the railroad also had its own physical impacts such as major shoreline changes along the rivers and lakes and substantial increases in fires due to sparks from the engines and the tracks.

The building of the Erie Canal and the consequent population growth along its route came with changes in the economy and even changes in the way ideas got communicated along its route. It also accelerated the rate of dispossession/removal of Native Americans in western New York and in the upper Midwest. The canal crossed the ancestral homelands of several tribes, including the Oneida, Onondaga, Cayuga, and Seneca, and pushed the indigenous people away from their land.

Change was inevitable with a venture as large in scale as the Erie Canal. The cliché about the half-full or half-empty glass is appropriate here. Depending on your vantage point, the canal was a blessing or a curse. At the time, celebrations over the new economic fortunes of the grand artificial waterway pretty much drowned out the negative views. Controlling the resource of water to change the fortunes of New York and beyond was a linchpin for modernization. In the early nineteenth century, the new waterway harkened to a prosperous future—nevertheless with blinders on.

Notes and Further Reading

Backman, Jr., Milton V. (1969) "Awakenings in the Burned-Over District: New Light on the Historical Setting of the First Vision," *BYU Studies Quarterly* 9 (Spring): 301–321.

Cross, Whitney R. (1982) *The Burned-over District: The Social and Intellectual History of Enthusiastic Religion in Western New York, 1800–1859* (Ithaca, NY: Cornell University Press).

Eisenstadt, Peter, ed. (2005) *The Encyclopedia of New York State* (Syracuse, NY: Syracuse University Press).

Klein, Milton M., ed. (2001) *The Empire State: A History of New York* (Ithaca, NY: Cornell University Press).

Koeppel, Gerard (2009) *Bond of Union: Building the Erie Canal and the American Empire* (Philadelphia, PA: Da Capo Press).

Shaw, Ronald E. (1966) *Erie Water West: A History of the Erie Canal, 1792–1854* (Lexington, KY: University Press of Kentucky).

Sheriff, Carol (1996) *The Artificial River: The Erie Canal and the Paradox of Progress, 1817–1862* (New York: Hill and Wang).

Stradling, David (2010) *The Nature of New York: An Environmental History of the Empire State* (Ithaca, NY: Cornell University Press).

"Two Hundred Years on the Eire Canal," (2019) *New York Heritage Digital Collections*, online, https://nyheritage.org/exhibits/two-hundred-years-erie-canal.

11 Building the Toronto Waterfront

The building of the Toronto waterfront beginning in the nineteenth century is an excellent example of shaping the land to utilize the coastal waters for the sake of commercial and urban growth. As environmental studies professor Gene Desfor and policy analyst Jennifer Laidley observed, "European settlement in North America began at the water's edge, where sheltered harbors offered protection for water-borne vessels essential for the basic needs of colonial expansion: defense and the movement of people, information, and commodities between empires and their outposts." With access to a network of waterways commerce expanded. Bigger vessels entering and leaving the harbor required "an infrastructure of wharves, quays, cranes, and yards that became a port, an area where ships could be loaded or unloaded." "Over time," they added, "ports took on a look of permanence" (Desfor and Laidley, 2011, 23).

Toronto has a long history of modifying its harbor and extending its shoreline which directly influenced the form, size, and scale of its city proper. Changes to the harbor in the nineteenth- and early-twentieth centuries were crucial to the city's development, along with its relationship to Lake Ontario, but with a cost.

Natural coastlines, in and of themselves, usually did not meet the requirements for water-borne commerce and security and had to be reworked to serve those ends. Decisions determining what and how to do it, how to fund it, and what priorities to give the port-building depended on many things, especially the prevailing landscape and the will and action of politicians and businessmen. Input from the people at large—for better or for worse—rarely played a role. No matter the intent, however, harbors and ports never stayed the same. They were altered by natural forces, such as tides, wind, and weather, and changed—sometimes dramatically—by human action.

Building all manner of structures along a harbor served a variety of purposes at different times in a port's history: forts with canon for protection, docks for boats, warehouses for storage, factories for production, and even parks for recreation. Rarely satisfied with the natural state of a harbor, city leaders across the globe sought ways to extend the shoreline along the edges of the town- or

DOI: 10.4324/9781003041627-16

cityscape and to turn marshland into "usable" land to create more space to make room for more people, businesses, and the various needs of the port.

Toronto is the capital of Ontario Province and the largest city in Canada with a metro population of over 6 million in 2020. About 5,000 years ago, settlements of hunting territories began to form, and people gathered at the mouths of the rivers to fish and trade. By 500 CE, there were 10,000 people in Southern Ontario comprised mostly of Algonquian speakers; by 1000 CE they moved into the region of present-day Toronto; by 1300 they had established villages there. Because of better farming land and warfare, Iroquoians began moving north from Toronto to join the Huron-Wendat Confederacy in Huronia. The confederacy collapsed or dispersed by 1650 because of inter-tribal warfare and disease brought by the Europeans. Algonquian speakers moved into the Toronto area about the same time. Of that group the Anishaabe established settlements in the area; some became known as the Mississauga. The Mississauga were dominant in the region until the end of the 1700s.

Since the early 1600s, French fur traders knew about a shortcut between Lake Ontario and Georgian Bay referred to as the *Toronto Passage*. The French built a small trading post on the Humber River (east of modern downtown Toronto) in 1720, but it failed financially and was abandoned in 1730. They then built a military (and trading) outpost called Fort Rouille (Fort Toronto) in 1750, east of the Humber. (In 1759 it was burned by the French garrison in retreat during the war with England). The site was chosen for its proximity to the harbor on Lake Ontario which is bordered by the Humber River to the east and the Don River to the west.

During the American Revolution, Loyalists moving northward into British-held territory settled along the upper St. Lawrence River and nearby lakes, resulting in the Province of Upper Canada (1791). After the war, Toronto (a centrally located town planned by the province's first Governor John Graves Simcoe) became an important site for the fur trade and for settlement. Concerned about possible threats from the young United States, the British decided that Toronto would be a good site for a naval and garrison base to control Lake Ontario and to aid in the movement of supplies and troops to the upper lakes such as Lake Erie. In 1793, a small town was laid out near the harbor called York (or Fort York after King George III's son the Duke of York), which would serve as the capital of Upper Canada. York is regarded as the birthplace of urban Toronto.

The name 'Toronto' was a Mohawk word—*tkaronto*—meaning "where there are trees standing in the water." It referred to the Narrows near present-day Orillia, where native groups created fish weirs. On French maps from the 1680s to the 1760s, the present-day Lake Simcoe was called *Lac de Taranto*. The spelling changed in the eighteenth century to Toronto, which referred to a large region that included the city of Toronto today. York is a former city within modern-day Toronto, southwest of North York and east of Etobicoke. As a separate city, it was one of six (including North York, East York, Scarborough, and Etobicoke) that was incorporated into Toronto in 1998.

First drawn up in 1787 and revised in 1805, the Toronto Purchase between local First Nations and the colonial authorities, resulted in the government controlling what is now Toronto, North York, Etobicoke, York, and Vaughn. In the early nineteenth century, York began attracting businesses and craftsmen and became a local market for agricultural goods. Its harbor was a particularly attractive feature, although by 1812 York had only 700 residents. Having manufacturing facilities on the lakeshore allowed for supplies to move to the inland more easily and for finished products to be exported efficiently.

Figure 11.1 Toronto harbor and Union Station looking east from John Street. Toronto, Canada, 1884. James Salmon. Public domain. Wikimedia.

During the War of 1812, York was raided twice by U.S. forces (requiring the garrison to be rebuilt in 1814) and retained strong antipathy for its neighbor to the south. York, nevertheless, continued to grow, becoming an important trading and banking center. Immigration from Great Britain and Ireland increased rapidly. In 1834, it was incorporated as the City of Toronto. In the 1850s, rail lines connected Toronto to New York and Montreal and across Upper Canada to Detroit and Chicago. In 1867, Toronto became the capital of the new Province of Ontario and rapidly was industrialized. Its economic growth was enhanced by railroad promotion, industrial tariff protection after 1879, and the opening of forests and mines to development in northern Ontario in the 1890s and 1900s. Hydroelectric power from Niagara Falls, beginning in 1911, provided cheap energy for factory growth. World War I expanded its financial and manufacturing influence.

Toronto is located on the shore plain alongside the harbor. Today it extends east and west of the harbor and far inland. The shore plain by the harbor has continued to be Toronto's downtown core. In 1793, Governor John Graves

Simcoe laid out a plain grid of streets along the eastern end of the harbor. As the town grew the basic grid pattern was extended, but ultimately gave way to less well-planned private development.

In 1816, construction of wooden wharves started along the harbor's north shore, although the shoreline itself remained sandy beaches. Storehouses, ship carpentries, and yards dotted the area. Economic ups and downs in the 1820s and late 1830s limited new wharf construction in several locations. An added problem was that the entrance to the harbor was getting smaller, including the advance of a shoal on the western channel. The colonial government appointed a committee to investigate issues of harbor maintenance and improvement in the 1830s.

In 1834, local authorities proposed a new pier (Queens Wharf) east of Garrison Creek and recommended that the Don River be redirected into the marsh, a breakwater be built at the tip of the peninsula near the harbor entrance to prevent silting, and asked the legislature to appoint a permanent harbor commission. But a controversy ensued over the committee's recommendations, and the colonial government was unwilling to authorize the proposed actions, although the plans for Queens Wharf would go forward. Local leaders would determine if anything else were to be done. But because Toronto's newly elected city council was bitterly divided between Reformers and Tories, nothing further happened.

Public investment in harbor improvement was important in Toronto, but debt undermined progress by the late 1840s. City merchants campaigned to have responsibility for management of the harbor handed to a trust, and in 1850 the provincial legislature created the Commissioners of the Harbour of Toronto commonly called the Harbour Trust. It was given responsibility for preparing plans and estimates for harbor improvements, managing the harbor, and regulating vessels there. Keeping the harbor entrance open was a constant concern requiring breakwaters and dredging.

Import/export activity continued to increase at the harbor during the 1840s. Any public consideration of maintenance and development issues or pressing concerns over contaminated water and sewage were ignored. From 1842 to about 1850, several new private piers along the north shore and retaining walls were constructed. As York University archivist Michael Moir stated, "This uncoordinated expansion of the port was driven by private commercial interests that began the city's push south into the bay in the search for additional space close to established businesses" (Desfor and Laidley, 2011, 29). Since the founding of Toronto, the port and lakefront had been contested space between public and private forces.

When the railways arrived in Toronto in the 1850s, they became formidable competition with the port by late in the nineteenth century. Initially, the harbor was an obvious terminus for railroad lines. This was particularly important with the completion of the Erie, St. Lawrence, and Welland canals transporting wheat, flour, lumber, and other bulk freight on the lakes and rivers connecting—directly or indirectly—to the harbor. The railways constructed a transport zone between the city and the lake. At that point, industrial areas developed on either end of the harbor along the rail lines. During the 1850s, the northern shoreline of Toronto Bay moved southward as the city gradually moved northward.

In the 1830s and 1840s, lack of available land along the waterfront limited economic growth, especially for shipping, industrial development, and railroad infrastructure. In the 1850s, a major campaign to promote lake-filling was underway to expand the shore south of the Esplanade. (Prior to this time private industry had engaged in making land periodically). What was intended to be a wide promenade adjacent to the lake, this public waterfront would be part of a beautification project to help clean up the shoreline. In the early 1850s, the construction of the Esplanade called for filling the waterfront, which would accommodate railroad tracks.

Ultimately, the Esplanade was taken over by the railways. In 1857, a bill passed granting the Grand Trunk Railway access to the Esplanade, suggesting that this action was in the best interest of the city. Over the next several decades, public reaction to the decision (and others dealing with railway access there) was acrimonious but did little to reverse the fate of the Esplanade. Only in the late twentieth century, when the railways abandoned their yards and shops in the area, was there a real opportunity to transform the expropriated public lands of the Esplanade.

According to historical geographer Thomas McIlwraith, "Had all filling been done within the most active construction period, 1852–1857, the picture of frenzied activity taxes the imagination: streams of wagons competing for space up and down the lakefront to maneuver into position for tipping spoil day after day, year in and year out" (McIlwraith, 1991, 15). Sources of fill included excavation from railroads and gravel pits, lakefront dredging material, ships' ballast, and wood for cribwork timbers. McIlwraith added, "For nearly two decades (particularly the late 1840s to the early 1860s) the Toronto waterfront had an emaciated appearance, pocked with little cesspools ... It was an ugly place, bearing the face of unbridled entrepreneurial ambition quite in contrast with the airy sward dreamed of by supporters of the promenade" (McIlwraith, 1991, 22). For the next one hundred years, the shore was extended to the south. (The original shoreline was north of the rail corridor). The filling continued until the 1950s when the modern shoreline was complete.

Toronto experienced tremendous growth in the 1870s due to industrialization, immigration, and suburban annexations. The 1880s, however, were the years that a variety of long-term plans to reconfigure the waterfront began to come together. The Trust's main project was the Commissioners' Cut (the Don Diversion), which would require a channel into the Don River then eventually into Lake Ontario. The work was only partially completed, and dredging continued to be necessary.

A variety of plans and schemes for the harbor emerged between the late 1880s and 1910 resulting in two common goals: the Ashbridge's Bay (sometimes referred to as Ashbridges Bay) marsh area located on the eastern flank of Toronto's harbor should be reclaimed for industrial purposes, and that waterfront lands should be held by a public authority.

In 1911, the Trust was replaced by the Toronto Harbour Commission. Its wide-ranging Waterfront Plan of 1912 was a decisive event in developing Toronto's waterfront, responsible for the shape of the modern shoreline. It called for extensive reclamation across the waterfront to create land for industrial, commercial, and recreational purposes.

A major focus was what to do due with Ashbridge's Bay. The marsh area there—created by the lower Don River as it empties into Toronto's inner harbor—included about 1,300 acres of land, marsh, and waterlots. As early as 1835 some had suggested reclaiming marshlands along the bay since silt from the Don River was regularly deposited there. Captain R.H. Bonnycastle argued that the marshes were "a fertile source of unhealthiness to the city" (Desfor, 1988, 88). Cholera scares were common there. The reclamation issue came up again in the 1850s, emphasizing the value of adding industrial facilities at Ashbridge's Bay. The use of reclaimed land for industrial purposes was further promoted in the late 1860s and early 1870s when the Harbour Trust created a breakwater for a navigable channel near the mouth of the Don River.

By the late 1880s, the breakwater along Ashbridge's Bay had caused stagnant water and sanitary problems. In addition, sewage and swill, slop and other waste from cow sheds (byres) were being dumped into the marshes. Little was resolved at the time, but ultimately a trunk sewer system to relieve the pollution was approved in 1908.

The 1912 plan called for turning Ashbridge's Bay Marsh into a huge new industrial district with waterfront parks and summer homes. To do so, the mouth of the Don was redirected into Keating Channel (concrete-lined) in 1914. Infilling of the marsh also began. By 1922, more than 500 acres of land was created on the former marsh, with another 500 due to follow. The lands were quickly industrialized, but the major waterfront park and adjoining houses were not constructed. Filling a large section of the harbor created a great deal of land for the city south of the Esplanade.

Plans on how to utilize the waterfront changed with the times. After World War II, suburbanization was increasing the region's population outside the city of Toronto. In 1951, the city itself had a population of 700,000 in a region of 1.1 million. In the 1950s and 1960s, planners began to consider newer and different ways to utilize the waterfront in a postindustrial era where industrial activity and port activity related to it were declining. A "new wave" of development was moving ahead for the waterfront in the early twenty-first century. A plan for revitalizing the waterfront was underway, initially stimulated by Toronto's efforts to host the Olympic Games. According to Desfor and Laidley in 2011, "'Mixed-use development' and 'sustainability' are the mantras of [a] thirty-year, multi-billion-dollar project, which promises to soften the waterfront's hard edges by 're-naturalizing' nature" (Desfor and Laidley, 2011, 13). Revitalization, of course, is an on-going process.

As Desfor, Laidley, and several other scholars and planners have made abundantly clear, Toronto's array of waterfront development efforts have had a vast impact on the local environment. Everything from extensive land reclamation to habitat destruction, from straightening and capturing the Don River to dumping wastes into the harbor have changed the waterfront in countless ways. This is a classic story of public versus private interests, economic livelihood versus a sustainable natural setting, experienced by cities throughout the world.

Using fill to increase land was a widely practiced enterprise globally in the nineteenth and twentieth centuries. The precise objectives may have been different, but the basic practice was the same, that is, to create wanted land out of unwanted land (or water). At one time, filling marshes and swamps to make "usable" space was viewed as a way to convert a noisome or useless site into solid ground to build upon. Extending fill land into oceans and rivers replaced what appeared to be an inexhaustible aquatic area with a more serviceable terrestrial one. There was little regard (or understanding) for the possible ecological impacts of creating new land, let alone declaring certain landscapes like marshes as worthless. But short-term goals trumped long-term consequences. Large parts of the Netherlands and the city of Rio de Janeiro, for example, are fill lands. Parts of Dublin, Saint Petersburg, Helsinki, Beirut, Mumbai, Shenzhen, Manila, Singapore, Montevideo, and Mexico City were constructed on human-made acreage. In North America, Montreal, New Orleans, San Francisco, Chicago, Boston, and Toronto, relied on the creation of fill land.

Today marshlands are recognized as important environmental resources. Native wetlands are flooded or saturated areas where water recurrently drenches the soil. They are a collection of wet environments that include salt marshes, inland freshwater marshes, wet meadows, bogs, swamps, and seasonally inundated floodplains. Ponds and shallow water zones of lakes also can be considered wetlands. These spaces provide several environmental, economic, and cultural benefits to proximate communities. Probably most significant is their role as water filters, thus maintaining surface-water quality in rivers, streams, and reservoirs—and even improving degraded water. Wetlands do this by retaining key nutrients, processing organic and chemical wastes, and reducing sentiment to receiving waterways. Wetlands also are important in flood control by storing water temporarily from overflowing riverbanks or collecting water in depressions. They are among the most productive and essential natural ecosystems in biomass production, which serves as food for small invertebrates, fish, other aquatic animals, and humans. Wetlands are key habitats, providing cover, breeding areas, and nurseries for all kinds of wildlife. As areas of biodiversity for plants and animals, they have few rivals.

Until late in the twentieth century, many people regarded marshes and swamps as worthless, bothersome, mosquito-infested, and unhealthy. Neither water nor land, they were seen as having value only when dried up, covered up, and turned into "usable" real estate. Without a good understanding of hydrology and ecology, early settlers saw no reason to treat marshlands differently than other lands they encountered.

A harbor and port are valuable assets for a city and nation. Especially before modern transportation and industrial practices, a harbor allowed access to the world and could make or break a coastal community. In a strange way, water was an impediment to economic development. Obviously, ships and other vessels depended on water connections, but without a port for picking up goods and dropping them off water-borne commerce was useless.

Toronto like many other cities was quick to see the advantages of developing their harbor, but for many reasons—most having to do with profound interests in economic growth. But often changes were made with little knowledge of the long-term impacts of their actions. Public versus private interests were wrapped up in the goals of growth, but often "public interest" was not easily defined.

Terraforming its shoreline went through many iterations in Toronto, depending on intent and practice. Sometimes the objectives went beyond what might be necessary or worthwhile; sometimes the aspirations did not connect to the reality of what repercussions change may have on people and the environment.

A Cree writer living in Toronto entered Coronation Park on the waterfront in July 2013: "So I move my shaker and sing: to the water, to the trees, to the earth, to the grandmothers and grandfathers of the four directions ... I sing for us all, so that we might find our own stories, here in the bright lights and gritty parks of the big city" (Methot, 2013).

Notes and Further Reading

Careless, James Maurice Stockford, Elaine Young, and Erin James-Abra (2018) "Toronto," November 6, *The Canadian Encyclopedia*, online https://thecanadianencyclopedia.ca/en/article/toronto.

Desfor, Gene (1988) "Planning Urban Waterfront Industrial Districts: Toronto's Ashbridge's Bay, 1889–1910," *Urban History Review* 17 (October): 76–91.

Desfor, Gene, and Jennifer Laidley, eds., (2011) *Reshaping Toronto's Waterfront* (Toronto: University of Toronto Press).

Goheen, Peter G. (2000) "The Struggle for Urban Public Space: Disposing of the Toronto Water in the Nineteenth Century," in Alexander B. Murphy and Douglas L. Johnson, eds., *Cultural Encounters with the Environment: Enduring and Evolving Geographic Themes* (Lanham, MD: Rowman & Littlefield Pubs.): 59–78.

Lemon, James (1989) "Plans for Early 20th-Century Toronto: Lost in Management," *Urban History Review* 18 (June): 10–31.

McIlwraith, Thomas (1991) "Digging Out and Filling In: Making Land on the Toronto Waterfront in the 1850s, *Urban History Review* 20 (June): 14–33.

Methot, Suzanne (2013) "Reclamation Project: A Cree Writer Reflects on Living in Toronto," *Canadian Geographic*, July 1, online, https://www.canadiangeographic.ca/article/reclamation-project.

Reeves, Wayne C. (1992) "Visions for the Metropolitan Toronto Waterfront, I: Toward Comprehensive Planning, 1852–1935," Center for Urban and Community Studies, December, online, https://tspace.library.utoronto.ca/bitstream/1807/91929/1/Visions%20Metropolitan%20Toronto%20Waterfront_TSpace.pdf.

12 The Lure of Falling Water: Niagara Falls

The lure of falling water. Niagara Falls might be the best example of that sensation in American History. The Falls are a natural wonder, a bare example of exploited nature, a widely recognized tourist attraction, and a source of massive hydropower. Geographer Patrick McGreevy asserted, "Few natural objects have attracted as much attention as the great cataract of Niagara. The contrasting realities of the Falls area—nature at its most sublime, popular recreation at its most tawdry, industrial might at its most power—have long competed to define Niagara in the popular imagination ..." (McGreevy, 2005, 1108). As a water resource, Niagara Falls has been contested space since the nineteenth century.

Niagara Falls demonstrated the great diversity of water use, from passive admiration of its beauty to aggressive manipulation of its power. The location of the Falls—bordering Canada and the United States—influenced its history in dramatic ways. When its initial inaccessibility was broached, the space around it (and how it would be used) changed in major ways. Niagara Falls is a physical place, a source of memories for those who viewed it, and a romantic subject of literature. To understand the importance of Niagara Falls is to explore its role as a natural spectacle, its attraction for tourists and gawkers alike, and its function as the first major source of hydroelectric power.

Niagara Falls formed more than 12,000 years ago at the end of the Ice Age when a deluge of water from melting ice drained into the Niagara River, which flows north from Lake Erie. (All outflow from the Great Lakes enters the Niagara River). The river runs through the Niagara Escarpment and connects to Lake Erie and Lake Ontario. Repeated glacial advances eroded away less-resistant rock leaving an escarpment (a long, steep slope) topped by hard dolostone and shale.

The tons of water from the river rushed over the edge of the Niagara Escarpment to what is currently Lewiston, New York. The water wore away rock which moved the Falls further upstream about seven miles and carved out Niagara Gorge. Before diversion for hydroelectric power, the volume of water over the large falls (or cataracts) exceeded 200,000 feet per second (or possibly 5.5 billion gallons per hour).

DOI: 10.4324/9781003041627-17

The Falls form the border between western New York State and eastern Ontario, Canada. They consist of American Falls (190 feet) and Bridal Veil Falls (181 feet) on the American side, and Horseshoe Falls (188 feet) on the Canadian side. They are the widest Falls in the world after Victoria Falls on the Zambezi River in southern Africa.

Indigenous peoples knew about the Falls for many years before European explorers. The first record of human settlers to the Niagara Peninsula came from southwestern Ontario between 1300 and 1400 CE. However, humans likely visited the area thousands of years earlier, including nomadic Clovis hunters. One of the earliest tribes was the Onguiaahra, from which the name "Niagara River" originated. Early settlers included an Iroquois group called Atiquandaronk—a name given to them by their enemies (Huron and Iroquois). They were called "Neutrals" by French explorers because they remained neutral in most conflicts between the Wendat (Huron) and the Haudenosaunee Confederacy (Iroquois). They lived in a region between the Grand and Niagara Rivers. A series of attacks and smallpox epidemics wiped out the population by the late seventeenth century. The Haudenosaunee continued to live in the area during battles for control among the French, British, and Americans. Today, Haudenosaunee remain in the Niagara Falls region either on reserves or in other communities.

Although there is no written record, the first European likely to visit the Falls was Etienne Brule, a French explorer, who lived among the Neutrals in 1626. He reported to his patron, the noted Samuel de Champlain, about the site. French explores from the St. Lawrence Valley engaged in the fur trade and missionary work as far west as Detroit in the early seventeenth century. The first eye-witness account of Niagara Falls among Europeans came from Father Louis Hennepin, a Belgian Recollet priest, who accompanied explorer and fur trader Robert de La Salle to the Falls in 1678. He recorded the account of that visit in *Description der la Louisiane* (a best seller) in 1683, paying more attention to the large numbers of Native Americans (ripe for conversion to Christianity) and the abundance of snakes than the prevailing stories and myths about the place or about the native culture itself. As popular historian Pierre Berton observed,

> Any Falls with which the Father was familiar would have been slender— long, lacy columns of tumbling water bouncing from crag to crag like some sure-footed alpine creature. But here there were no mountains, and that astonished and puzzled Hennepin. The river coursed across a flat, forested plain and then, without warning, split in two and hurled itself over a dizzy cliff. "I could not conceive," he wrote, "how it came to pass that four great Lakes ... should empty themselves at this Great Fall, and yet not drown a good part of America."
>
> (Berton, 1997, 15)

For Hennepin, Niagara Falls was "the most Beautiful and at the same time most Frightful Cascade in the World." This dichotomy in perception lived on for years, while the Falls largely remained a remote place through much of the eighteenth century (Berton, 1997, 18–19).

The area around the Falls became a French stronghold, but highly disputed. The location was based on the need to defend trading interests around the Great Lakes from the expansion of the American colonies. Forts were built on the mouth of the Niagara River, intending to control access to the Great Lakes. The French built the first fort above Niagara Falls in 1679 (Fort Conti) but it was replaced in 1687 by Fort Denonville. It only lasted about a year. The first permanent fort, Fort Niagara, was constructed in 1726.

The period of French control of the area ended in 1759 with the British capture of the fort and the defeat of the French elsewhere in Canada. The British settled west of the Niagara River and eventually elsewhere in southern Ontario. Agricultural settlement west of the river proved to be an important food supply line for the British military during the American Revolutionary War. When the war ended in 1783, the border between the United States and what became Ontario was established as the middle of the Niagara River. The Falls, nonetheless, remained inaccessible to all but a few visitors at the time.

Accessibility—and massive changes that went with it—took time and effort. Not until the early 1800s was basic access infrastructure in place. Between the late 1700s and the middle of the 1800s, boats were the primary means of getting to the Falls. The first human-powered passenger ferry opened in 1820 to carry people across the Niagara Gorge. In the 1820s, a stairway was built down the bank at Table Rock and the first ferry service began across the lower Niagara River. In 1827, a paved road extended from the ferry landing to the top of the bank on the Canadian side (which later became a prime location for hotel development).

At this time, construction of roads and turnpikes in western New York opened approaches to Niagara Falls. Gravel tolls roads connected the Falls to Albany 300 miles away. Most important was the completion of the Erie Canal in 1825 providing water transportation from eastern New York to nearby Buffalo, from there allowing ordinary travelers to approach the Falls by carriage or ferries.

The 1840s and 1850s saw changes to transportation in and out of the Niagara Falls area that would transform the site of sublime visages and natural wonder into a tourist trap and industrial center. Between 1849 and 1962, thirteen bridges were built across the Niagara River Gorge, although few remain. In 1848, under the direction of Charles Ellet, Jr., the first service bridge across the Niagara Gorge was completed. Although a bridge could have been built more easily a few miles downriver, the growing fame of the Falls led engineers (and investors) to select a site near the cataract. The bridge, however, never served railroad traffic.

Railroads in Canada and the United States especially encouraged greater access to the Falls. By 1840, Buffalo had two daily steam-powered trains connecting to there. By the 1890s, ninety-two daily trains made stops at Niagara Falls during the summer. In 1855, John August Roebling of Brooklyn Bridge fame built the Niagara Railway Suspension Bridge, the first bridge of its type.

It had two levels—one for carriages and another for railway traffic. Historian William Irwin stated, "It is easy to see why a bridge in front of Niagara Falls would attract attention to American engineering. Just as the Falls symbolized America's nearly boundless and unadorned nature, so bridges embodied Americans' eager extension of progress and civilization over the wild landscape" (Irwin, 1996, 32).

Figure 12.1 Prospect Point, Niagara Falls. The booth advertising "Photographic and Stereoscopic Views of the Falls" belonged to American photographer Platt D Babbitt, who held the monopoly on views from this vantage point (1859). William England. Public domain. Wikimedia.

Some natural wonders in North America escaped the kind of exploitation that Niagara Falls experienced, but the Falls quickly became an example of how to drag down such a unique setting. By the mid-nineteenth century, the days had passed when the few visitors were lucky enough to view the Falls in their essentially pristine state. Charles Dickens visited Niagara Falls during his trip to the United States and Canada in 1842. In an April 26 letter, he said of

Horseshoe Falls, "It would be hard for a man to stand nearer God than he does there." His visit to the Falls shaped his later writing, especially his water imagery (McKnight, 2009, 69).

The extent to which tourism led to tawdry displays of commercialism—on both sides of the border—was stunning. Initially, the Ontario side of the Falls was basically rural compared with emergence of the New York side as a lively commercial and recreational center, but that would change. With their control of Canada, the British at first prohibited private development on the northern side of the Falls when it declared property along the riverbank a Crown military reserve in 1780. In the 1790s, tops of trees and bushes were cut off to view the cataract. On the American side, New York State owned a one-mile strip of land around the Falls but provided no tourist facilities. After the sale of public lands on both sides in the nineteenth century the landscape was forever changed.

Tourism at the Falls began in the 1820s and increased tenfold within fifty years, becoming the area's dominant industry. The opening of the Erie Canal coincided with the rise of the commercial and industrial middle class especially in the United States, which helped to open the tourist floodgates. The advent of the railroads in the 1840s only accelerated the flow of tourism. The number of visitors to the Falls leaped from 20,000 in 1838 to 45,00 in 1847. By the Civil War, the increasing amount of attention to mass entertainment at the Falls diminished the site as an elite resort as greater numbers of working-class people discovered the attraction. The automobile further democratized Niagara Falls tourism. It was not because the Falls became accessible to all people that tourism sought the lowest common denominator of entertainment, but because developers sought economic gain any way they could and as quickly as they could but did little to honor that transcendent place.

In the 1820s local landowners began to change the landscape around the Falls specifically to accommodate the flood of visitors: stairways, paths, footbridges; a stone tower with a panoramic view; hotels, museums, and peddlers of all kinds of merchandise. Private developers bought up the best overlooks, fenced them off, and charged high prices. As early as the 1830s married couples began to arrive at the Falls to celebrate their honeymoon, largely as a result of promoters urging "honeymooning" as a tradition in the mid-nineteenth century.

In 1846, one of the most famous attractions at Niagara Falls, *the Maid of the Mist*, had its initial voyage as a ferry, charging a fee to transport people and cargo across the river. When a bridge across the gorge eroded business, *the Maid of the Mist* became a sightseeing boat, taking visitors close to Horseshoe Falls. The *Gazetteer of the State of New York* noted in 1860 that Niagara Falls had available hotel space more than other locations of equal size in North America.

The singular splendor of the Falls as a breathtaking site was replaced by an almost circus-like atmosphere of souvenir shops, amusement parks, and sideshows. In more recent decades, the Canadian side of the Falls grabbed a greater share of the tourist market, with Ontario sporting its own gaudy attractions and even a gambling casino.

Daredevil feats, from tightrope walking to sending barrels over the Falls, gain wide publicity. The first known stunt was arranged by William Forsyth of the Pavilion Hotel in 1827. He decorated a boat as a pirate ship—populated it with a bison, two bears, two raccoons, a dog, and a goose—and sent it over the waterfalls. The bears escaped, but the other animals perished. In 1859 "the Great Blondin" (Jean-Francois Gravelet) was the first to walk across 1600 feet above the gorge on a 1,100-foot-long tightrope. Annie Edson Taylor, a Michigan schoolteacher, was the first person to go over the Falls in a barrel in 1901 and she survived. She was sixty-three years old at the time, but claimed to be forty-two, and saw the stunt as a way to make money. Not everyone was as lucky as Taylor, but she never made the fortune she anticipated, and instead worked as a Niagara street vendor for twenty years, then died penniless. Although the Ontario government made stunting within the park boundaries illegal in 1951, the practice did not end.

Despite the explosion of tourism and activities that seemed to trivialize the area, there was interest in setting aside outdoor recreational space near Niagara Falls to help preserve some of the natural landscape from commercial and industrial development. In 1869, the builders of famed Central Park in New York City, Frederick Law Olmsted and Calvert Vaux, and Buffalo lawyer William Dorsheimer, called for returning the area to a more natural state. They wanted public purchase of land along the Niagara River, but the response was slow. In 1880, the New York State Survey issued a report, "Preservation of the Scenery of Niagara Falls," which made the case for the state to purchase land and to begin a restoration effort. It took until 1883 for Governor Grover Cleveland to sign a bill creating the Niagara Falls State Reservation. The state spent $1.5 million to purchase the land through eminent domain, and it opened in July 1885 as the first state park in the United States. It encompassed 435 acres (296 of which are under water). By 1906, more than 150 privately owned businesses were removed from the area.

In 1885, the Province of Ontario established the Niagara Parks Commission, beginning the process which led to Queen Victoria Park opposite Horseshoe Falls. Like its American counterpart, Queen Victoria Park was meant to recover a good deal of land despoiled by crass tourism and industrial activity and was originally part of the upper Niagara Riverbed. The park officially opened to the public on May 24, 1888—the birthday of Queen Victoria. The original park site was 154 acres between present-day Clifton Hill and Cynthia Islands (Dufferin Islands). Several private properties were expropriated to make way for the park.

Unlike the Niagara Falls State Reservation, Queen Victoria Park received no public monies for management and maintenance. Its commissioners faced the problem of operating a park that was self-supporting but free to visitors. As a result, concessions were rented to private businesses within the park grounds. The distance between the parklands and tourist village was so great that it called for a roadway to join them, given wide access to trains and eventually automobiles. Power operations of three companies sat on some of the prime land in the park by the early 1900s. Yet, as Irwin argued, everything was not perfect at the Niagara Falls Reservation either: "Niagara's most insistent boosters decried the pestilential convergence of pickpockets, bunco men, gamblers, dope fiends, prostitutes, strippers, and ruffians on the Riverway just beyond the reservation ..." In addition, he added, "Ominously ... the establishment of the reservation opened the floodgates for power interests and manufacturers to exploit the areas outside the park" (Irwin, 1996, 208, 210).

Any attempt to coordinate the interconnection of the two parks never transpired. Olmsted was frustrated that planting nonindigenous plants and trees at Queen Victoria Park was given priority over developing a natural setting. Nevertheless, Queen Victoria Park was an impressive site. The *Toronto Evening Telegram* remarked "To the visitor who has not seen the Falls for four or five years, the change in the scene would be most striking." Gone were "regular gulling establishments, where exorbitant prices were charged for insignificant articles to unsuspecting foreigners and other visitors ..." "But the voice of the fakir," it proclaimed, "is heard no more ..." Instead, it was replaced by "the broad acres of the new park, tastefully laid off into walks and drives and planted at intervals with young trees ..." (Berton, 1997, 149).

Despite efforts to "clean up" the area around Niagara Falls, signs of a tourist economy did not disappear over the years. The enticement of falling and running water also had attracted industrial growth alongside the tourist business in the nineteenth century. Waterpower had been a staple of economic development for generations, and the Niagara River and then the Falls were obvious sources to tap. First came water mills and then came hydroelectric power.

The French were the first to build mills near the rapids of the Niagara River, although the idea of harnessing power for industrial use in the area went back to the eighteenth century. State policies encouraged industrial development, and soon land along the river's banks was auctioned off for industrial use. Given the success of the Erie Canal and others, entrepreneurs proposed a system of canals for the area. It was not until 1861 that the first canal-feeding water mills were opened by the Niagara Falls Hydraulic Power & Manufacturing Company.

From about 1875 to 1900, the Niagara Falls area became a successful industrial site with power derived from the Niagara River and only modestly from the Falls. A concentrated manufacturing district existed at the top of the gorge downstream. Local mills manufactured paper, rubber, plastics, petrochemicals, abrasives, and metallurgical products. The Niagara became one of the largest chemical producers—with consequent water and air pollution—in the world. This industrial development lasted into the 1960s.

As early as 1805 settlers on the American side of Niagara Falls were enthusiastic about the possibilities of industrialism by tapping the Niagara's power. Just before the War of 1812, the first village at Niagara Falls was named "Manchester" after the great industrial center in England.

As McGreevy stated, "Although the village's name later was changed to Niagara Falls, the dream of an industrial metropolis remained because people continued to imagine that Niagara's power was unlimited." By the 1890s, various plans had been proposed for larger power development harnessing the Falls' power, and, according to McGreevy, "All were tinted with an image of Niagara Falls as an inexhaustible, almost magical source of power" (McGreevy, 1987, 51).

It took several years before the dream of utilizing the waters of Niagara Falls for large-scale power development came to pass. The first successful attempt to use water from the Falls took place in 1875 by cutting a canal in the plateau which was level with the top of the Falls. Water was let into the canal above the falls and emptied over an embankment into the river. Mills used waterwheels to capture the water, which supplied power to a flour mill, paper mill, barrel factory, brewery, and a fork and spoon factory. There also was a small generator that lit arc lamps. The first central power station in Niagara's milling district began operations in 1882, but it was just a start. Most mills in the area used waterpower from the rapids for their operations, not from the cataract.

In 1892, entrepreneur William T. Love came to Niagara Falls with a dream to create a model city—a utopia—which would be built around cheap waterpower abundant in the area. The plan centered on diverting water from the upper Niagara River through a power canal to generate power. He envisioned a city of 200,000. His venture, however, went bankrupt in 1896.

The so-called Evershed Plan was the first significant scheme to exploit the vast power of the Falls for large-scale use. A division engineer on the Erie Canal, Thomas Evershed, had suggested a plan for 200,000 horsepower from Niagara Falls in 1886. It called for a systematic use of the falling water by building a series of inlet canals to serve hundreds of mills, while a huge runoff tunnel would extend under the town of Niagara Falls and empty water back into the river. Niagara River Hydraulic Tunnel, Power & Sewer Company (Niagara Falls Power Company) obtained a charter from the state in 1886 to divert water from the river outside the Niagara Reservation. Funding was difficult and between 1886 and 1890 a series of problems stymied construction of the great tunnel, plus existing waterwheels or turbines could not withstand the great pressure of the Falls. In addition, no effective way to transmit power existed at the time, especially since there was not enough local need to justify production of so much power.

Only in 1889 did the Niagara venture have sufficient financial support to attempt to overcome the problems faced by the initial Evershed Plan. The Cataract Construction Corporation, a holding company incorporated in New Jersey which purchased all the stock of the Niagara Falls Power Company, offered enough capital to move forward. Among the investors were bankers J.P. Morgan, John Jacob Astor, and several Vanderbilts. There remained great risk in the venture since technical methods to achieve the desired ends remained unproven.

Ultimately using the Falls to produce electrical power (initially for lighting only) became the focus of the project. The big question was how to produce and transmit the power and by what means. Engineers agreed that electrical power could be produced by using waterwheels or turbines. (The Swiss had successfully manufactured equipment for such purposes). The village of Niagara Falls was too small to become the main market for the electricity, so a more distant site became necessary. Buffalo (twenty miles away) was the logical site but determining how to get the power to the city was a knotty problem. Transmitting such a large amount of power had never been done before.

Attention was now focused on whether to employ direct current (DC) or alternating current (AC) for the transmission system, giving attention not only to technical merits but to economy as well. A debate over the virtue of Thomas Edison's direct current system (used in the first commercial power plant in the United States at Pearl Street Station in New York City in 1882) versus long-distance transmission with alternating current championed by inventor and businessman George Westinghouse already had been raging. The "battle of the currents" was a key confrontation in the fledgling electrical power industry that fundamentally transformed the world of power and light.

In a country still largely decentralized, the DC system offered little hope of bridging the miles. The low-voltage DC systems had distance limitations because they experienced "voltage drop"—the further the current traveled, the larger the voltage loss on the line. While Westinghouse did not have expertise in electricity, he learned to rely on bright young inventors like the eccentric Nikola Tesla and in buying up potentially useful patents. The key to AC power was the transformer which reduced high-transmission voltage to safe levels by stepping down voltages at substations along the line. In 1886, Westinghouse Electric Company built the first commercial AC system in Buffalo.

The field of power transmission was so new that the Cataract Construction Corporation decided to consult several authorities. Among the first was the celebrated Edison, who promoted his own DC system. In the end, they chose Westinghouse's AC system over Edison's DC system and over an AC system proposed by General Electric. Westinghouse had great momentum not only with his Buffalo project but also for getting the lighting contract for the Chicago World's Fair in 1893 (by underbidding Edison). His system illuminated the "White City" at the centennial with some 8,000 arc lights and 130,000 incandescent lights and ran machinery in a dramatic display of AC's potential.

In the same year, Westinghouse was awarded the contract for the alternators and transformers to power the proposed hydroelectric plant at Niagara Falls. Tesla developed the polyphase induction motor, which used multiple AC currents to produce rotation. The induction motor and its electrical system allowed long-distance application of electricity. General Electric was given the contract for building transmission lines from Niagara Falls to Buffalo.

In 1895, the Niagara power station went into service with the first of three 5,000-horsepower AC generators (eventually eight). George Forbes, designer of the Niagara plant, described it as "one of the greatest engineering works in the world" (Melosi, 2008, 94–95). On August 26, 1895 the first power was delivered to the Pittsburgh Reduction Company (now Alcoa). A transmission line to Buffalo was completed the next year.

In 1895, the awesome power of falling water at Niagara Falls was successfully harnessed and converted into electricity for commercial and industrial uses. The engineering work at Niagara Falls in the 1890s played a commanding role in moving from the mechanical power of the waterwheel to the production of hydroelectricity. In turn, the role that the generation of hydroelectricity played was vital in the process of electrification itself and in the long-distance transmission of electrical power. Electricity, which had been until recently a curiosity, was a harbinger of a new age. Historian David E. Nye observed, "While electricity had been a key element in the Columbian Exposition, the Buffalo Pan-American Exposition in 1910 made electricity its major theme...For Buffalo, progress and the future were intimately linked to electrification, and it was logical to make electrical generation the fair's major theme" (Nye, 1997, 41–42). The same could be said for other cities aspiring to grasp modernity.

At the turn of the new century, the electric power from the Falls was being utilized exclusively for American lighting and for American industry. The market in Canada much smaller at the time, and the government had shown modest interest in developing power on its own. Steam power still had the attention of many Canadian businessmen. In April 1900 the Toronto Board of Trade established a committee to get information about electricity but concluded that it would remain "a secondary force, a handmaid or servant of steam or some other primary power" (Berton, 1997, 207–208).

Americans actively held back electrical power production in Canada in the 1890s, retaining exclusive rights to develop electricity from the Canadian falls. The Canadian Niagara Power Company was owned wholly by Americans. In 1903, the Niagara Falls Power Company got the Canadian and provincial governments to construct a power plant inside Queen Victoria Park. Canadian Niagara Power Company, a subsidiary, would supply power in New York State and in Ontario (first power delivered in 1905).

In addition, a major coal famine in Ontario in 1902 pushed the local government to consider a more activist role on Canada's part in exploiting the Falls for electrical power. Soon municipal efforts pressured government to question the actions of private power companies operating in Canada, hoping to reverse privatization of hydroelectric development. The debate continued for several years.

As competition for hydropower increased along the border the level of water diversion from the Falls for power production became a serious concern for both sides. In 1906 the Burton Act limited Niagara diversions, calling for a treaty to preserve scenic beauty. In 1909 the Boundary Waters

Figure 12.2 Save Niagara Falls - from This (1906) by J.S. Pughe (1870–1909). Public domain.
    Wikimedia.

Treaty created the International Joint Commission and further limited diversions. At the time most of the electricity produced by the diversions was
exported to the U.S. American producers endorsed diversion limits because
this would entrench their water rights. However, during World War I all
diversion limits were lifted.

  The Federal Water Power Act in 1920 initially was meant to coordinate the
development of hydroelectric projects in the United States and moved diversion limits to those set by the Boundary Waters Treaty. The act itself was
modified several times over the years, and efforts to develop compatible use of

the Falls between the two countries continued through much of the early twentieth century.

Under a key treaty between the two counties in 1950, the majority of the Niagara River was diverted to enormous hydroelectric complexes. According to Daniel McFarlane, historian and a leading expert on Niagara Falls technology, "In the early Cold War period, the United States and Canada cooperatively remade Niagara Falls by building massive new hydroelectric complexes and water control works. The manipulation of this famed landmark blended the organic and the mechanical, leading one member of the public to label it 'a completely artificial and man-made cataract' in reference to the extent to which the Horseshoe Falls ... had become part of the built environment" (McFarlane, 2020, 109).

At Niagara Falls the power plant also became part of the tourist scene. Once open, guided tours took visitors inside. Writer Ginger Strand observed, "Some visitors liked the power plant more than the waterfall. H.G. Wells wrote an article for *Harper's* after his visit in 1906. 'The real interest of Niagara for me was not the waterfall, but the human accumulations about it,' he declared. 'The dynamos and turbines of the Niagara Falls Power Company impressed me far more than the Cave of the Winds'" (Strand, 2008, 167).

Nye discussed the ritualized illumination of Niagara Falls with the advent of electrical lighting which "recontextualized and revisualized" the Falls, impressing viewers "simultaneously with the awesomeness and the beauty of a natural scene and with the skill and power of those who 'recreated' it" (Nye, 1997, 59). Yet questions remained as to whether the technical advances represented by hydropower production at Niagara Falls were harnessing nature or vandalizing it.

Tourism and power generation converged at Niagara Falls. In a piece published in 1895, geologist G.K. Gilbert wrote:

> The great cataract is the embodiment of power. In every second, unceasingly, seven thousand tons of water leap from a cliff one hundred and sixty feet high, and the continuous blow they strike makes the earth tremble. It is a spectacle of great beauty ... Its charms are the theme of many gifted bard and artist, but the fascination of its ever-varied yet continuous motion, and the awe that waxes rather than wanes with familiarity, are not to be felt at second-hand; and so the world, in ling procession, goes to see. Among the multitude there are some whose appreciation of its power has a utilitarian phase, so that they think most of the myriad wheels of industry its energy may some day turn; and there are a few who recognize it as a great natural engine, and in its activity and its surroundings see an impressive object lesson of geographic progress.
>
> (Gilbert, 1895, 203–204)

The falling water of Niagara Falls was so enticing that over its human-contact history a tug-of-war between the Falls' aesthetics and its utility never stopped raging. Admiring the sublime Falls led to garish tourism and then to a cluster of nearby manufacturing. The ultimate act of exploitation was developing a massive hydroelectric system. Niagara Falls became a hybrid of the power of Nature and the power of power. As McFarlane suggested, "While there had been considerable public pressure throughout the first half of the twentieth century to preserve the scenic beauty of Niagara Falls, the Canadian and American governments generally privileged hydroelectric development and only did what was necessary to appease appearance advocates" (McFarlane, 2013, 779). Such was the lure of falling water.

Further Reading

Berton, Pierre (1997) *Niagara: History of the Falls* (New York: Kodansha International).

Gilbert, G.K. (1895) "Niagara Falls and Their History," *National Geographic Monographs* (New York: American Book Company, September).

Irwin, William (1996) *The New Niagara: Tourism, Technology, and the Landscape of Niagara Falls, 1776–1917* (University Park, PA: Pennsylvania State University Press).

McFarlane, Daniel (2013) "A Completely Man-Made and Artificial Cataract": The Transnational Manipulation of Niagara Falls," *Environmental History* 18 (October): 759–784.

McFarlane, Daniel (2020) "Nature Empowered: Hydraulic Models and the Engineering of Niagara Falls, *Technology and Culture* 61 (January): 109–143.

McGreevy, Patrick (1987) "Imagining the Future at Niagara Falls," Annals of the Association of American Geographers 77, 48–62.

McGreevy, Patrick (2005) "Niagara Falls," in Peter Eisenstadt, ed., *The Encyclopedia of New York State* (Syracuse, NY: Syracuse University Press), 1108–1110.

McKnight, "Natalie (2009) Dickens, Niagara Falls and the Watery Sublime," *Dickens Quarterly* 26 (June): 69–78.

Melosi, Martin V. (2008) *Thomas A. Edison and the Modernization of America* (New York: Pearson).

Nye, David E. (1997) *Electrifying America: Social Meanings of a New Technology* (Cambridge, MA; MIT Press).

Strand, Ginger (2008) *Inventing Niagara: Beauty, Power, and Lies* (New York: Simon & Schuster).

Tinkler, Keith (1987) "Niagara Falls 1750–1845: The Idea of a History and the History of an Idea," *Geomorphology* 1, 69–85.

Part V

The Mid-Twentieth Century

The water episodes in this section may not appear to have anything in common, but they each deal with how to use water to meet specific ends in the mid-twentieth century. In doing so, unanticipated consequences were ignored or sidelined. This disparate grouping deals with the wide physical imprint of the Houston Ship Channel along the coast of the Gulf of Mexico; the chronic flooding of the Mississippi River which Louisianans (and others) hoped to contain with a "Levees-Only" strategy; and contention between salmon fishing and the generation of hydropower along the Fraser River in Canada and what that meant for Canadian–American relations.

When it opened in 1914—the same year as the Panama Canal—the Houston Ship Channel (HSC) was a means to elevate Gulf Coast commerce and trade to new heights and place the City of Houston at the center of this economic boon. Those goals were accomplished, but it is important to understand that the ship channel was more than an effective access point to the open sea; it was the center of an industrial region stretching from Buffalo Bayou in Houston to the Gulf of Mexico. Like the Erie Canal in New York, such a massive physical presence changed southeastern Texas in major ways. The HSC's environmental (ecological) footprint is massive. The World Wildlife Federation provides a good definition of an environmental footprint, calling it:

> ... the impact of human activities measured in terms of the area of biologically productive land and water required to produce the goods consumed and to assimilate the wastes generated. More simply, it is the amount of the environment necessary to produce the goods and services necessary to support a particular lifestyle.

Viewing the HSC in this way goes well beyond the specific site, measuring the resources necessary for the channel to function as an economic engine for the region, but also taking into account what this industrial activity has done to the air, land, and water surrounding it.

The flooding problem on the Mississippi River can be compared to flooding problems around the world. In this case, the proposed

DOI: 10.4324/9781003041627-18

solution—"levees only"—raises several questions. First, to what degree was this policy based on solid scientific and technological theory, or was it simply part of a long-standing tradition believed to have worked in the past? The practice of building levees had an extensive past in Louisiana, but also was the focal point among competing ideas, some of which acquired political favor beyond their scientific authenticity. Second, who did "levees only" benefit? This approach to flood control did not take into sufficient account the hydrological and environmental knowledge about the need for and effects of flooding. Instead, human needs—especially those of small farmers, plantation owners, and city dwellers—mattered. And in the hierarchical society of the South, the impact of flooding on those who were marginalized (the landless poor and people of color) did not matter much at all.

The story of salmon and the Fraser River in British Columbia is an interesting one. It combines the fate of the wild salmon population, growing demand for hydroelectric power, and the tensions associated with transboundary issues. The case, however, is not so much a classic example of the preservation of a crucial fish species versus the voracious desire to develop new sources of energy for a modern, industrializing world. It is a story of how Canadians and Americans attempted to sustain two different types of resources for economic use. In the end, the Fraser River avoided hydroelectric development and continued to be a source of salmon (although not to the same degree as in the past), while energy companies went elsewhere, especially along the Columbia River.

13 The Houston Ship Channel's Environmental Footprint

The Houston Ship Channel (HSC) is a vast waterway—a combination of numerous natural features and extensive human construction and dredging. It overwhelms the landscape as a major transportation route, a massive industrial corridor dominated by petroleum refining and petrochemical production, and a central component of the Houston metropolitan region. There is little doubt that the Houston Ship Channel since its completion in 1914 had a major economic impact on the city and the related industries, but it also is the core of an environmental footprint extending from its origins in Buffalo Bayou to the Gulf of Mexico. What happens in the ship channel impacts everything around it. As such, it is difficult to think of the ship channel simply as just another waterway. It is a human-made structure that has changed the ecology of the region from the City of Houston into Galveston Bay and the Gulf of Mexico.

The Houston Ship Channel in southeast Texas is one of the busiest water courses in the United States. At the turn of the twenty-first century, it was the leading U.S. port in foreign tonnage and second in total tonnage, ranking as the sixth-largest port in the world. In 2003, 6,300 ships passed through its waters. The waterway originates in the City of Houston along Buffalo Bayou to the Turning Basin, where the actual ship channel begins, which is 52 miles from the Gulf of Mexico. It then heads east and south through a series of bays meeting several rivers and bayous, then spills into Galveston Bay at Barbour's Cut. The ship channel is wedged between Bolivar Peninsula and Galveston, heading out to the ship anchorages in the Gulf of Mexico.

The idea for a ship channel can be traced to Houston merchants in the 1850s who were dissatisfied with those who ran the Port of Galveston (about 50 miles south of Houston) which controlled access to the Gulf of Mexico and to the shipping lanes beyond it. These merchants envisioned a route that would bypass Galveston and they began to initiate their plans after the Civil War. It would take time to overcome Galveston's dominance. By 1900, it ranked first in the United States in exporting cotton and third in wheat. Its momentum to become an international port of note was obstructed by the Great Hurricane of 1900 and railroad development around Houston—major reasons that Galveston lost its status to the rival city.

DOI: 10.4324/9781003041627-19

The rise of shipping and commercial development along Buffalo Bayou in the nineteenth century was an essential precedent to the construction of a ship channel. The navigable Buffalo Bayou flows for 65 miles west to east from Katy in Fort Bend County to a tributary of the San Jacinto River at Lynchburg. The river continues into Galveston Bay. In 1826, John Richardson Harris established the town of Harrisburg at the confluence of Buffalo and Brays Bayous (just below the present-day Turning Basin) and set up the first industry (a steam sawmill) on Buffalo Bayou. The new town served as a port of entry and trading center for early settlers in the region prior to the Texas Revolution. On April 19, 1836, two days before the Battle of San Jacinto which led to Texas Independence, the Mexican army looted and burned Harrisburg.

Persuaded that Harrisburg was an attractive location for a port, but unable to acquire property in the vicinity, Augustus Chapman Allen and John Kirby Allen decided to settle approximately 15 miles upstream at the confluence of Buffalo and White Oak Bayous. While the brothers' glowing claims about the wonders of the townsite (named after the Texas Revolution's hero, Sam Houston) were clearly overstated, they persuaded the Texas Congress to designate it as the temporary capital of the new republic.

Houston ceased to be the republic's capital after 1839, but a prosperous and productive shipping industry and commercial trade was established there along the shores of Buffalo Bayou with local, regional, national, and international impacts. "The commerce of the republic [of Texas] very largely followed the flows of nature downriver to the Gulf" (Meinig, 1969, 59). Galveston Bay provided the best access to the rich agricultural lands to its northeast, and with few rivers navigable for long stretches, access to those lands via Buffalo Bayou made Houston the state's chief inland port. Barges and riverboats from Buffalo Bayou could load cargoes onto seagoing ships in Galveston and, in turn, pick up goods to deliver into the hinterland.

Initially swampy and overgrown with vegetation, Buffalo Bayou required clearing many snags and fallen trees before it became a significant water highway for shallow-draft steamboats running between Houston and Galveston. It soon was viewed as an important access route for the inland cotton trade, as well as an important point of transshipment for a wide variety of goods that entered or left the state. It also was an avenue of passage for travelers and immigrants.

The small steamboat *Laura* was the first to dock at Houston's port in 1837. In a bit of hype, Buffalo Bayou was designated as the "National Highway of the Republic" in 1840. The next year, the Houston City Council established the Port of Houston and levied wharfage fees to help finance dredging and port improvements.

Not only was Buffalo Bayou's advantage over other local waterways that it ran east and west, but it also was relatively wide and deep at its mouth along the San Jacinto River. At the entrance to Galveston Channel, vessels had to traverse a 12-foot bar, and then run over a shell reef stretching across the middle of Galveston Bay. It was not uncommon for ships to run aground at this juncture as they traveled toward Buffalo Bayou. Where the waters of the San Jacinto River entered Galveston Bay ships faced another bar. It became obvious that the connection of the natural waterways along the route from Galveston Bay to Buffalo Bayou would have to be "improved" to allow for effective trade to prosper.

In 1869, the Buffalo Bayou Ship Channel Company initiated major dredging and widening of the bayou. The first federal survey for a ship channel running from Buffalo Bayou to Galveston Bay (1871) stated that the bayou was at least 70 feet wide and could be navigated to Houston by vessels drawing less than four feet. Some stretches had been much deeper (15–20 feet). As a way of avoiding the high port charges at Galveston, the Houston Direct Navigation Company (chartered in 1866) loaded and unloaded ocean vessels in the channel and carried the cargoes by barge along the bayou. Soon Galveston wharves were being bypassed altogether.

Cotton from the west plantations found its way to the Houston port, as did sugar, cattle, and other commodities from the rich Brazos agricultural region. In the nineteenth century, "cotton was king" in Texas, and in the 1870s Houston was the center for exporting cotton to textile mills in the northeastern United States and Great Britain. The bayou also served as a source of power for grist mills and sawmills. Not unlike cotton, Houston was "the capital of the Texas lumber industry." By 1875, the ship channel along the east stretch of the bayou was widened and straightened, allowing most ocean-going vessels to deliver goods to the Houston area.

Charles Morgan, who was involved in Gulf shipping and had been eager to bypass Galveston's wharfage costs, bought the Bayou Ship Channel Company in 1874. Within two years, he had dredged a channel from Galveston Bay to present-day Clinton (southeast of the Turning Basin). He also stretched a chain across the channel at Morgan's Point to collect tolls, making Buffalo Bayou a nautical tollway. "In response to local protest," historian David McComb stated, "the United States government agreed to purchase Morgan's improvements in order to liberate the stream" (McComb, 1981, 65). Morgan soon turned to railroading.

Through the efforts of men such as Houston Congressman Thomas H. Ball (a member of the Rivers and Harbors Committee) appropriations became available to expand upon Morgan's channel improvements. This was particularly important since Galveston was developing plans for deepening the bar to allow ocean-going vessels access to the local wharves. Congress authorized the channel to be deepened to 25 feet and a terminus to be located at Long Reach, later called the Turning Basin. By 1909, the channel had only been deepened to 18 ½ feet.

A delegation from Houston traveled to Washington urging Congress to accept the "Houston Plan" for the channel, which would provide half of the cost of the remaining dredging. Congress quickly accepted it with assurances that the facilities would remain in public hands; the Texas legislature enabled Harris County (where most of Houston resided) to establish a navigation district; and the citizens of the county approved a bond issue. The work began in 1912. In September 1914, the dredging was completed on the Houston Ship Channel, and the "port that built a city" was officially opened on November 10 with a big celebration and a parade in downtown Houston.

Deep-water capacity was delayed until after World War I. In 1919, the *Merry Mount* became the first ocean-going vessel to ship cotton directly from Houston to Europe. Within a decade, Houston was the leading cotton port in the United States, matching its role as the largest spot cotton market in the world and the second leading port in the country in the volume of cotton orders handled.

Oil, however, would soon rival cotton as the Houston Ship Channel's most important cargo. The ship channel became the focal point of a vast center for petroleum refining and petrochemical production which took advantage of one of the world's great concentration of oil fields after the 1901 Spindletop strike in nearby Beaumont. Initially, oil was transported to the channel by barge but eventually through a vast pipeline network.

Refineries (and later petrochemical facilities) concentrated along the Houston Ship Channel, and by the 1930s Houston was on the map as the "energy capital of the world." At that time, half of the world's production of oil was located within 600 miles of Houston, and it could boast of 4,200 miles of pipeline reaching outward to hundreds of fields. While Beaumont was close to the major oil fields, it did not have the railroads, banking system, or port facilities that Houston had already developed.

The utilization of Buffalo Bayou as a major commercial conduit for cotton, timber, and other commodities established an important precedent that allowed the Port of Houston to build an industrial capacity, which would surpass its nineteenth- and early-twentieth-century enterprises. In addition, before the dominance of oil and gas, Houston and the Gulf Coast also was the site of other extractive enterprises beyond timber, namely sulfur, salt, lime, and other minerals. However, oil, natural gas, and petrochemicals would lead the Gulf Coast trade and its industrial production for years to come.

In a reciprocal way, transporting and refining oil and other petroleum-related goods added incentive for additional channel improvements. The oil and gas industry created thousands of jobs in production, transportation, sales, refining, and distribution; it led a variety of companies to establish headquarters in Houston and the surrounding area; it invested in local infrastructure; and it became home to many oil-related businesses such as Hughes Tool Company and Cameron Iron Works. The oil and gas industry also was a central reason for Houston's ascendancy as a major city.

Between 1929 and 1945, oil and related industries replaced cotton as the central feature of the Houston economy. In 1935, almost half of all Texas oil was shipped through the Port of Houston. Historian Lynn Alperin stated, "Buffalo Bayou has been transformed from a meandering stream into a vast industrial complex" (Alperin, 1977, 114). *Fortune* magazine concluded that "without oil Houston would have been just another cotton town."

Although all elements of the industry were important, according to oil historian Joseph Pratt, "refining left its mark most prominently on the region." While the Gulf Coast refining region, writ large, extended from New Orleans to Corpus Christi, the "historical and geographic center" was focused along a 100-mile coastline from Houston to Port Arthur, which contained the greatest concentration of refineries. The large, integrated oil companies—several of which were based in Houston—were most influential in shaping refining development along the Gulf Coast (Pratt, 1980, 3–7).

With the construction of pipelines from a variety of oil fields in Texas, Louisiana, and Oklahoma, oil continued to flow to the Gulf Coast despite the subsiding of the Spindletop boom. (The first strike there was in 1901.) With the end of World War I, refineries slowly began to be attracted to the area, but it was in the 1920s that the largest plants were constructed in earnest.

Emerging as a major economic force in World War II, the production of petrochemicals added to the importance of petroleum and natural gas to the Houston area in general and to the ship channel in particular.

World War II brought significant changes in the industry, especially with the need for aviation fuel, synthetic rubber, and other petroleum-based products necessary for the war effort. In fact, half of the synthetic rubber used in the war came from Texas. By 1950, there were 27 chemical plants along the ship channel. By the 1980s, the Houston area had more than half of the petrochemical capacity in the country.

High levels of air, land, and water pollution accompanying the economic bounty is the price the region has paid for the concentration of petroleum and petrochemical facilities and other industries and the extensive urban growth of Houston. Some of the environmental challenges begin in the ship channel but include a variety of activity all along the urban/industrial corridor from Houston to the Gulf of Mexico.

There are many bayous (slow-moving streams) that drain Houston. Early in the city's history, the bayous served as a water supply for humans and livestock and for fire protection but soon proved to be too unsanitary for drinking. They have carried rainwater, runoff of all kinds, dead animals, and sewage long before the completion of the ship channel, but urban growth and industrial development after 1914 made conditions worse.

By the late nineteenth century, the wanton pollution of the bayous highlighted the need for adequate sewerage, ultimately resulting in a modest underground sewer system in the early twentieth century. McComb stated, "In 1945, after complaints by residents and a poliomyelitis scare, the U.S. Public Health Service inspected the malodorous Brays Bayou and found flowing into the small stream enough raw sewage to equal that produced by a town of 54,000 people." Further investigation at the time into the bayous found that Buffalo Bayou water was 80 percent sewage and other waste (McComb, 1981, 147). Wastewater collection and treatment, however, remained uncoordinated in Houston and Harris County for many more years.

To this day, the bayous carry the bulk of rainwater for the city and have become the focus of chronic flooding—including the Houston Ship Channel. Beginning in the late 1930s, the primary emphasis of the city's flood-control plan was building Barker and Addicks Reservoirs (to the northwest of downtown) and more than 2,500 miles of cement-lined channels built along the bayous by the U.S. Army Corps of Engineers between the 1940s and 1970s. Canalization proved to exacerbate flooding as residential and commercial development to the west of the city and the laying of extensive nonpermeable surfaces contributed to more runoff more quickly entering the bayous than could be managed. Increasingly serious rain events only added to the problem. The flood waters, of course, make their way to the ship channel, creating siltation problems there, and also carry highly polluted water to Galveston Bay.

In 1967, federal investigators called the pollution problem in the Houston Ship Channel "overwhelming" (Melosi and Pratt, 2007, 40–41). A member of the Federal Water Control Administration stated after a 1967 inspection that "The Houston Ship Channel in all frankness is one of the worst polluted bodies of water in the nation. In fact, on almost any given day this channel may be the most badly polluted body of water in the world. Most days it would top the list."

The water in the channel contains innumerable chemicals, grease, and debris from the ships and the various industrial processes. As ship traffic got greater, so did the number of oil spills, explosions, and collisions between tankers, freighters, and barges in and near the HSC. Heavy rains that occasionally flushed the channel caused, among other things, fish kills in the bay.

Because of the heavy industrial activity, air quality—particularly for nearby neighborhoods—often was compromised. In the 1950s, a study found that air pollution in the ship channel area had high concentrations of sulfur dioxide and hydrogen sulfide along with higher-than-average dust, chlorine, sulfate, and nitrogen dioxide (Melosi and Pratt, 2007, 73). Making the various problems worse was the indifference and lack of attention to the pollution over the years until at least the 1970s (McComb, 1981, 148–149).

In some cases, criticism over the polluted water was squelched. In 1920, the director of the Port of Houston, Colonel Benjamin Allin, took action against the Galena Signal Oil Company for discharging effluent into the HSC. Despite support from the mayor of Houston, Allin barely avoided dismissal after the Galena attorney protested his actions (Gorman, 2001, 18–19). As historian Hugh Gorman stated, "If a 'water pollution' equivalent of Los Angeles smog had existed, the Houston Ship Channel probably would have been involved. Few places in the nation had as great a concentration of industrial plants disposing such a wide variety of wastes into a single channel" (Gorman, 2001, 242). Numerous hazardous waste sites—many eventually designated as Superfund sites—dot the land along the ship channel, the offcasts from oil and petrochemical plants and other industrial properties (Melosi and Pratt, 2007, 28–33, 37, 52–53, 56–57, 260–273).

Possibly the most serious pollutant found in the channel was poly-chlorinated biphenyls (PCBs, first manufactured in 1929). They were linked to various forms of cancer in lab animals and possibly humans. Although PCB production was banned in 1979, they remained in the bottom of the channel and in many fish species. Periodically, advisories limiting or ceasing to eat fish contaminated with PCBs (and dioxins) from the channel were issued along the Texas coast. Historian Jack E. Davis dramatically stated, "After Spindletop, [the ship channel] resolved into a maritime highway for the petrochemical industry, with a slurry of heavy metals, dioxins, and PCBs to name a few—all headed for Galveston Bay" (Davis, 2017, 419).

A range of other problems dogged the HSC. During hurricane season, there always was a chance that storm surge would create oil spills and cause major damage to storage tanks in the channel or even worse. Subsidence, which occurred when substantial amounts of groundwater was pumped from aqui-fers, was a chronic problem in the Houston area (a city extremely dependent on groundwater at least until the mid- to late-twentieth century). Along the ship channel alone subsidence measured up to ten feet by 1978 (Melosi and Pratt, 2007, 130–131).

Pollution control for the ship channel, the whole industrial corridor, and certainly the greater Gulf Coast region itself was a massive undertaking. As Pratt argued, "Those who called for stricter controls of pollution had to overcome more than regional attitudes favoring growth. Local politics, as well as civic leadership, were dominated by business leaders whose idea of a 'healthy business climate' included low taxes, weak unions, and very limited regulation" (Melosi and Pratt, 2007, 9). Industries along the HSC preferred self-regulation with respect to pollution rather than government-imposed sanctions. Efforts at regulation on the state and federal level met with modest success at best. But anti-pollution laws were a long time in coming, and we still await very stringent ones.

A practical challenge for the Houston Ship Channel and the surrounding area was the need for constant vigilance to keep the waterway operating. This meant dredging:

> Dredging is an activity that is required to be carried out to remove the unwanted deposits from water pathways. But even though the activity aids regularity in marine traffic, it is not without its disadvantages. Dredging possess a huge threat to the marine environment and is required to be carried out quite carefully aided only with the help of the right dredgers and dredges.
>
> (Chopra, 2019)

Early dredging activity was undertaken either with little concern or little awareness of its environmental impact, which could alter the composition of the soil deposits in the channels and thus disturb existing marine habitats. Turbidness in the water could alter the soil composition and possibly create newer and harmful organisms. Ultimately, toxic particles might spread and further pollute the water.

Another concern was the size and location of spoil disposal areas created by dredging. Early on, spoil was side cast in open water or placed in proximity to the dredge area along inland streams. Before World War II, dredged material within the interior of Galveston Bay was deposited in open water. Since approximately the 1960s specific regions were designated as disposal areas. Between World War II and 1993, there were about 27,000 acres of such sites in the Galveston Bay system (including the upper Houston Ship Channel) (Ward, 1993, 151).

While dredging along Buffalo Bayou was underway in the late nineteenth century, the U.S. Engineer Department worked to link Galveston and Houston's channel through dredging in the bays and rivers. (There would be a 32-mile-long channel that connected the Houston Ship Channel to the Gulf of Mexico.) By the 1880s progress in dredging was inadequate and continued to be so into the new century.

In 1912, the Army Corps awarded a contract to the Atlantic, Gulf, and Pacific Company (Atlantic) to complete a full 52-mile, 25-foot-deep channel in 3½ years. Siltation in the channel proved to be a serious impediment. Even though the Houston Ship Channel opened in 1914 (to the authorized depth of 25 feet), Atlantic continued to work on it for over 50 years. By 1961, it had dredged more than 153,000,000 cubic meters of material. Seven main Houston Ship Channel deepening projects took place between 1899 and 1998. Dredging activity continued in and around the channel into the twenty-first century. Much of the maintenance work in the HSC was performed by government dredge activities, while deepening work was typically privately contracted.

Ships enter and exit the Houston Ship Channel via Galveston Bay. This is the largest bay in Texas covering approximately 600 square miles. The Galveston Bay watershed is about 2,400 square miles extending from metropolitan Houston north along the Trinity River basin and past the Dallas-Fort Worth Metroplex. Galveston Bay is an estuary, that is, a partially closed coastal body of water connected to the sea through the Gulf of Mexico. Within it, fresh water from the San Jacinto and Trinity Rivers and the bayous and creeks from Houston and surrounding areas mixes with salt water. The average depth is eight feet—accounting for the need to dredge channels connecting to the ship channel. In addition, as Courtney Smith stated, "Inflows of fresh water from rivers, bayous, and streams are the lifeblood of an estuary, bringing in nutrients that fuel the food chain and sediments to replenish our wetlands. Because of this, estuaries are among the most productive ecosystems in the world…Ninety-five percent of commercially and recreationally fisheries species in the Gulf of

Mexico are dependent upon estuaries like Galveston Bay during some part of their life cycle..." (Smith, 2013, 9)

The importance of Galveston Bay in providing shipping access in and out of the HSC is significant. However, there is competition for the use of its waters from commercial fishers, shrimpers, and oystermen; from recreational fishers and pleasure boat operators; from vacationers and permanent residents along the coastline; and from offshore oil rig operators. Such varied and extensive uses of the bay all contributed to or were affected by changes to the estuary.

Human impacts on the Texas coast, especially Galveston Bay, have been wide-ranging. As noted earlier, extensive dredging of channels and passes resulted in the discharge of sediment into the bay, ultimately modifying natural water circulation patterns, changing salinity, and impacting water quality and estuarine plants and animals. Because of increased cultivation, the construction of irrigation and drainage canals, and urban paving resulted in many streams accelerating the transport of sediment and nonpoint pollutants (including pesticides and herbicides). Between 1951 and 1970, a substantial portion of the estuary was closed to shellfish harvesting because of bacterial pollution from raw and unchlorinated sewage (Smith, 1972, 21–22).

Also, thermal pollution from various manufacturing processes and power generation can be lethal to fish. The discharge of organic materials, trace metals, and so forth too numerous to mention from a variety of sources—including oil production, pipelines, spills, and chemical production—added significantly to the pollution load of all water courses. Human actions have had significant impacts on the major habitats of the bays and environs, including oyster reefs, submerged aquatic vegetation, intertidal marsh vegetation and animal life, and fresh water wetlands. For example, between 1950 and 1989, approximately 54 percent of fresh water marshes in the Galveston Bay watershed were lost because of draining wetlands and conversion to upland areas (Lester and Gonzalez, 2001, 12). Oil leaks from offshore drilling became a more recent concern. Hurricanes and sea-level rise have added to major impacts on the gulf.

After the construction of the Houston Ship Channel, people began to raise concerns about beach contamination, the oily water, declines in fishing, and other manifestations of pollution in and around the bay. Local governments and the state had incentives for pollution control, especially given the various groups contesting the use of Galveston Bay. However, industrial interests took precedence at least until the late twentieth century, especially with promises of self-regulation. Until federal water and air pollution laws became more stringent beginning in the 1960s, there were no major regulations that changed the course of Galveston Bay's pollution problems. The 1924 Oil Pollution Act, which prohibited ships from flushing their tanks within three miles of shore, proved relatively weak, although it put some pressure on refineries to improve their operations.

In 1987, the Galveston Bay Foundation was established. The intent of the nonprofit corporation was to address issues and concerns of a variety of bay

Figure 13.1 Houston, Texas (Dec. 11)—The Port of Houston is accessed by a 54-mile long ship channel. Here, a ship passes under the Fred Hartman Bridge on the Houston Ship Channel, December 11, 2000. USCG photo by PA2 James Dillard, United States Coast Guard. Public domain. Wikimedia.

interests, and, among other things, published key reports germane to conditions in the gulf. Other groups, such as the Galveston Bay Estuary Program and the Galveston Bay Information Center identified problems in the bay and provided data about the state of the bay to aid in protection and management.

Not all the environmental problems that Galveston Bay is facing can be traced to the Houston Ship Channel and to Buffalo Bayou upstream from it. However, many of the bay's environmental problems demonstrate the interconnection of Buffalo Bayou, the ship channel, and the bay, and graphically show how the pieces are connected and interdependent. It is, therefore, too simplistic to think of the HSC—despite its enormous economic impact—as an isolated environmental sacrifice zone.

One study referred to the ship channel simply as "a ditch through a bay, a river and a bayou" (Vincent et al., 2015, 469). But it is much more. It is collectively a group of waterbodies, a huge industrial corridor, and part of an urban system operating interactively. According to one study, it operates within "networked ecologies" requiring that we study its various components interconnected in a single landscape (Holmes, 2013, 3). This perspective effectively shows the enduring importance of the Houston Ship Channel, not simply as a key industrial waterway, but as part of a whole region with a significant environmental footprint. Once again, a decision based on economic opportunity—such as building the ship channel—can threaten the very environment in which it was meant to thrive. The environmental (or ecological) footprint is not just a geographic designation, but an area of real and potential risk to that opportunity and to all of us.

Notes and Further Reading

Alperin, Lynn M. (1977) *Custodians of the Coast: History of the United States Army Engineers at Galveston* (Galveston: Galveston District, United States Army Corps of Engineers).

Chopra, Karen (2019) "Effects of Dredging on the Marine Environment," *Marine Environment*, online, https://www.marineinsight.com/environment/effects-of-dredging-on-the-marine-environment/#:~:text=Effects%20of%20Dredging%20on%20the %20Marine%20Environment.%20Dredging,the%20help%20of%20the%20right%20dredgers%20and%20dredges.

Davis, Jack E. (2017) *The Gulf: The Making of An American Sea* (New York: Liveright Pub.).

Gorman, Hugh S. (2001) *Redefining Efficiency: Pollution Concerns, Regulatory Mechanisms, and Technological Change in the U.S. Petroleum Industry* (Akron, OH: The University of Akron Press).

Holmes, Rob (2013) "Houston Ship Channel," *Dredge Research Collaborative*, online, http://dredgeresearchcollaborative.org/works/houston-ship-channel/.

Johnson, Glenn S., Steven C. Washington, Denae W. King, and Jamila M. Gomez (2014) "Air Quality and Health Issues Along Houston's Ship Channel: An Exploratory Environmental Justice Analysis of a Vulnerable Community (Pleasantville)," *Race, Gender & Class* 21: 273–303.

Lester, Jim and Lisa Gonzalez, eds. (2001) *Ebb & Flow: Galveston Bay Characterization Highlights* (Galveston: Galveston Bay Estuary Program).

McComb, David G. (1981) *Houston: A History* (Austin, TX: University of Texas Press).

Meinig, D.W. (1969) *Imperial Texas: An Interpretive Essay in Cultural Geography* (Austin: University of Texas Press).

Melosi, Martin V. and Joseph A. Pratt, eds. (2007) *Energy Metropolis: An Environmental History of Houston and the Gulf Coast* (Pittsburgh, PA: University of Pittsburgh Press).

Pendergrass, Bonnie B. and Lee F. Pendergrass (1990) *In the Era of Limits: A Galveston District History Update, 1976-1986* (Washington DC: U.S. Army Corps of Engineers).

Pratt, Joseph A. (1980) *The Growth of a Refining Region* (Greenwich, CT: JAI Press, Inc.).

Sibley, Marilyn McAdams (1968) *The Port of Houston: A History* (Austin, TX: University of Texas Press, 1968).

Smith, Courtney (2013) "Galveston Bay: A Brief History of One of America's Great Waters," *Houston History* 10: 8–13.

Smith, James Noel (1972) *The Decline of Galveston Bay* (Washington, DC: The Conservation Foundation).

Vincent, Mark, et al. (2015) "The History of Dredging at the Port of Houston: Ditching High and Low to Build a Port," *Proceedings of the Western Dredging Association and Texas A&M University Center for Dredging Studies*, 469–486.

Ward, George H. (1993) *Dredge and Fill Activities in Galveston Bay* (Galveston Bay National Estuary Program).

14 "Levees-Only" in Louisiana and the Great Mississippi Flood

One of the fundamental impacts of too much water out of place is flooding. Over the course of human history, flooding has been viewed as an inevitable part of life that must be tolerated (and often feared), or a force to be controlled and even utilized in some fashion. As one headline read: "Floods Can Be Inconvenient. Large Floods Can Be Downright Disastrous."

A classic case in the United States was the Mississippi River flood of 1927, the attempted control measures leading up to it, and its repercussions with a focus on the State of Louisiana and the commitment to a "levees-only" approach to flood control. The story of the 1927 flood has been measured against all such inundations that preceded it and followed it. The Mississippi flood also demonstrated the difficulty and complexity in finding ways to anticipate, live with, or manage such events. Flooding may be an equal opportunity disaster or biased toward one group of people or another, or one location or another. It comes in many forms and many contexts.

The North American watershed system is an awesome force. The Mississippi Basin alone drains more than 40 percent of the United States' land, including parts or all of 31 states as well as two Canadian provinces, covering 1,245,000 square miles from the Appalachian Mountains in the East to the Rockies in the West. It extends approximately 2,340 miles from the headwaters at Lake Itasca, Minnesota, to the Gulf of Mexico. The Mississippi flows through one of the flattest areas in North America, but its channel is irregular and constantly in flux, which complicates navigation and flood control. It also is alluvial, that is, sediment-carrying, with extensive fluctuation in volume. The Mississippi Delta is an expansive sedimentary deposit at the mouth of the river, which makes up one-third of the total area of Louisiana.

Going back to the colonial era, the great rivers were regarded as untapped or underutilized resources, raw material waiting to be harnessed, managed, and exploited for human benefit. Efforts to control the Mississippi River in Louisiana began in the eighteenth century. The primary technology employed was the levee—an extended naturally occurring ridge or an artificially constructed (mostly earthen) wall or fill meant to regulate water levels in rivers and streams. Especially on the lower Mississippi, levees were designed to keep

DOI: 10.4324/9781003041627-20

the river in its banks and thus reduce the threat of flooding to protect nearby farmland and human settlements.

The French built the first levees (doubling as roadways) on the Mississippi in 1717 to protect Louisiana's major city, New Orleans. In 1722, construction began on a four-foot earthen levee on top of the natural levee in the Crescent City. By 1727, it extended about one mile along the waterfront. Other levees were already in place at plantations upriver from the city. French and Spanish law stated that a landowner must agree to build a levee—of specific height—on his property before obtaining legal possession. Most of these levees were four- to six-feet high and usually built 30–40 yards from the natural banks. According to historian George S. Pabis, "Near-sighted economic interests motivated planters to build levees hastily, blocking outlets and draining swamps without concern for the long-term consequences of their actions" (Colten, 2000, 65).

By 1724, however, levee construction may have raised spring flood levels to a new high and contributed to flooding in New Orleans. Further flooding in the city prompted the Superior Council to require all settlers from the mouth of the river to Cannes Brulees (a 50-mile stretch) to build levees. By 1732, a levee system was completed from 12 miles below the city to 30 miles above on both banks. By 1752, the network extended about ten miles more. The levees were uneven in height, consisting of both natural levees and artificial structures, and most likely had gaps. Privately built levees varied in design and effectiveness. Flooding along parts of the route affected some farmers more than others.

This levee system to protect New Orleans was completed by the end of the first French period in Louisiana. Levees served as protection from flooding in and outside the city to be sure but also were used for drainage, irrigation, and fertilization in the surrounding agricultural area. Since the colonial levees were earthen structures that had been reinforced with timber, they leaked. During floods, water oozed through crevasses (breaks) into drainage ditches that flowed to cypress swamps, then drained back into the river downstream. The seepage could be used for irrigation or replenished with fertile silt. The levees thus changed the landscape of farms and plantations in major ways.

Generally, the best farmland available was along the riverbank, which was higher and drier than other parts of the property. Deeper in a lot the land could be wetter and not well drained, thus yielding fewer crops. Areas in the back of lots, where water often accumulated had to be drained and cleared, adding new dikes to protect the low-lying areas. To make way for new dikes built on top of natural levees, farmers moved buildings and structures away from the water. Movement back from the river often increased the risk of inundation, which led to even higher levees. Slave labor, once concentrating on planting and harvesting crops, now was being utilized also for clearing timber and working on the levees.

The range of activities that were changing farms and plantations along the river—including transforming cypress swamps, savannahs, and natural levees, building levees and ditches, changing cane breaks into cattle pastures, and cultivating rice—fundamentally altered the area's environment and put them at odds with the natural flow and function of the Mississippi River.

The U.S. Territory of Louisiana passed its first levee law in 1807, giving authority for maintaining levees to the parishes, which were responsible for holding rural landowners accountable for levee building and maintenance. When Louisiana became a state in 1812, levees extended from about the mouth of the Red River above New Orleans to below the city on its east bank. Yet, there was no uniform design or no central flood protection authority, and thus levees remained a fragmentary barrier and thus limited in effectiveness. New Orleans, by contrast, continued to maintain its municipal levees after Louisiana became a U.S. possession. By the early 1800s, levees were regarded as essential to life along the Mississippi, but they did not eliminate flood hazards. Indeed, they often contributed to redirecting risk to open floodplains and elevated the volume of water to greater heights raising the flood level.

In the 1780s and the 1790s, the Spanish and Americans constructed canals to shorten transportation on the river. These cutoffs, it was argued, would reduce transportation time on the river and would hasten water to the Gulf of Mexico before it could back up and overflow. The possible risk would be to redirect water (and flooding) to other locations such as the Atchafalaya Basin in south-central Louisiana. Canal proposals, along with endless discussions about levees, complicated developing a sound flood management system for many years.

With the end of the War of 1812, Americans began a more aggressive plan to drain swamps, close natural outlets on the Mississippi, and build more levees in Louisiana. Floods, such as the 1828 flood, spurred another rigorous levee-building operation. New Orleans and a variety of parishes along the river passed legislation regulating levee dimensions, raised taxes for new construction, and established rules for maintenance. By the 1840s, levees extended irregularly from New Orleans to the mouth of the Ohio River. Yet, a flood in 1844 inundated plantations in Arkansas, Mississippi, and Louisiana.

In the late 1840s, several engineers working for the Louisiana state government alerted the legislature about the failure of levees meant to protect people living along the Mississippi. Some of the engineers supported relying on both outlets and levees for flood protection, criticizing the current practice of depending solely on levees. They argued that confining the river within a single channel forced sediment to settle on the riverbed, thus raising the level of floods, rather than collecting on alluvial lands to create more topsoil. Proponents of a "levees-only" approach, such as engineer Caleb G. Forshey, believed that the levees would confine the river to a single channel and thus force the river to create a deeper channel for and by itself.

Plantations built near the riverbanks, destruction of marshland, and more levees were leading to damaging floods—and more levees. The flood in

1849, in particular, was devastating—possibly the worst flood in the nineteenth century. It continued for 48 days, with one in ten (or about 12,000) residents of New Orleans having to leave their homes temporarily. The river was particularly high below the mouth of the Red River, bringing water from new plantation districts along the Red, Ouchita, Tensas, and Yazoo Rivers (Morris, 2012, 106–107, 119). The *New Orleans Daily Crescent* observed about flooded cemeteries, "The water comes on, and covers the houses of the dead as well as those of the living. Like Time, in its ravages, the river makes no distinctions..."

Flooding in 1849–1850 inundated large sections of New Orleans and the Delta farmlands. The devastating flood inspired intense debate on controlling flooding on the Mississippi. Louisianans also had complained for years that the federal government had not reimbursed them or the state for building levees and clearing rivers that led to the reclamation of thousands of acres of public lands. Louisianans claimed that in 1829 there was approximately 5.4 million acres of swamp land in the state, of which, they asserted, 2.9 million acres was reclaimed by 1849. (Future environmental debates would raise serious questions about reclaiming swamps and marshland.)

The Swamp Land Acts of 1849 (applied only to Louisiana) and 1850 were the first major efforts on the national level to confront flood control in the Mississippi Valley. Their intent was to encourage states to act. Under their provisions, the U.S. government transferred about 28 million acres of unsold swamp land—unfit for cultivation—to state governments to sell to private buyers. The rationale came from earlier policies to finance canals and railroads. Revenue from the sales was to be used to fund flood control and wetland drainage.

Flood control was regarded as secondary to new settlement and farming. The cost of reclamation was very high. The work in Louisiana especially was mismanaged. Since the law did not adequately define "unfit for cultivation" of these lands, privatizing this property often led to abuse. Almost 65 million acres of land was given away based on false claims of worthlessness. By the mid-1850s, the program had largely failed (Andrews, 2006, 97–98; Cowdrey, 1977, 9; Morris, 2012, 106–107, 140–141).

Despite the failure of the Swamp Land Acts, the federal government was now involved in flood management, especially through the U.S. Army Corps of Engineers. The Corps was created in the late eighteenth century to construct military fortifications, but it emerged in the nineteenth century as a source of engineering expertise in numerous areas, especially navigation. It conducted topographic surveys, and in the 1820s completed surveys on the feasibility of river and harbor improvements. Under the Rivers and Harbors Act of 1824, it was responsible for constructing, among other things, water resource development projects for interstate commerce and economic development. In a landmark case, *Gibbons v. Ogden*, the Supreme Court decided that federal authority over navigable rivers was a matter of interstate commerce, and thus justified the Army Corps' survey activities (Andrews, 2006, 89, 95).

In 1850, Forshey drew up the report of the Senate Standing Committee on Levees and Drainage, which was the first major study of the Mississippi River commissioned by the state. It called for levee districts, formulated levee dimensions, provided a survey of all the levees on the Mississippi and a survey of the Atchafalaya Basin. Forshey's conclusions differed from those of the committee, especially his staunch belief in levees-only. His survey of the levee system was important nonetheless because it called for a statewide levee system at the time (to be completed in three years) when levees in most parishes demanded a great deal of work. There were approximately 847 miles of levees along the river, and Forshey calculated that another 159 miles were needed. Crevasses occurred in every major flood, but engineers had no precise evidence of their effect on the river (Colten, 2000, 67–68).

It was difficult for the State of Louisiana to solve its flood problems on its own, and engineers exhorted the federal government to provide more information about the river. In 1850, Secretary of War Charles Magill Conrad started in motion what became the Mississippi Delta Survey, initially under the direction of Captain Andrew A. Humphreys and Lieutenant Colonel Stephen H. Long. Humphreys was, at the time, the most famous hydrographer in the country. Under the direction of the secretary of war, a second survey was undertaken by well-known private civil engineer Charles Ellet, Jr. Ellet's brief study, *The Mississippi and Ohio Rivers,* was published in 1851. Humphreys took ill in 1851, and his survey work ceased until 1857. Humphreys and Henry L. Abbot (in charge of field operations after 1857) published their massive study, *Report on the Physics and Hydraulics of the Mississippi River,* in 1861 (Colten, 2000, 70–71).

In his report, Ellet called for levees being extended, but also recommended new outlets as well as reservoirs constructed at the Mississippi's headwaters to guarantee an important flood-control precaution for the lower parts of the river. No action was taken on Ellet's recommendations. When Humphreys and Abbot's report became public in 1857, it overshadowed Ellet's approach and became, as Army Corps historian Martin Reuss argued, "the most significant contribution that the Army engineers made to hydraulic engineering in the nineteenth century" (Reuss, 1998, 50).

While the theories that underpinned the 600-page study later proved flawed, the report was very thorough and so full of new data that it remained influential for years, receiving national and international recognition. At its heart was the notion that "levees-only" could control flooding on the lower Mississippi River. In so doing, it rejected Ellet's ideas about reservoirs on the headwaters to help control flooding. It also argued that in most cases cutoffs and outlets were "too costly and too dangerous." Humphreys' plan gained wide support from southern politicians and engineers, who viewed it as making practical sense, and sustained the view of riparian owners that levees were a tradition worth continuing. Also, building levees appeared to be the answer to reclaiming more land rather than investing in outlets and diversions that were costly and used up valuable farmland (Reuss, 1998, 50–52).

The coming of the Civil War turned attention away from less immediate concerns such as river flooding. Humphreys' report did not have impact on the ongoing engineering debate until the 1870s. Soon after the war, Secretary of War Edwin M. Stanton ordered Humphreys to tour the levee system demolished during the conflict. He reported that in the Delta alone there were 59 crevasses, one extending two miles long and flooding thousands of acres. Humphreys' report added that some sort of federal authority was necessary to build the mainline levees, but little was done as the nation took steps to re-unite. Resources in Louisiana were exhausted at this time, and Washington D.C. provided the only hope for financing the rebuilding of the levees.

Opposition to federal flood control came from the Reconstruction North and also from politicians seeking to limit federal authority in a number of areas. Louisiana tried to make some repairs, but its resources were extremely limited. The Panic of 1873—a financial crisis that triggered an economic depression in North America and Europe—made matters worse, as did a disastrous Mississippi Valley flood in 1874.

In the wake of the flood, Congress created another board, the Levee Commission, to come up with a permanent plan for reclamation of the alluvial basin of the Mississippi. Its 1875 report was notable for finding 143.4 miles of crevasses south of Commerce, Missouri; dreadfully poor and badly run levee boards; and 107.5 miles of levee destroyed in Louisiana alone since the end of the war (Cowdrey, 1977, 23–24).

The commission fully supported the Humphreys-Abbot report which stated that cutoffs, outlets, reservoirs, and diversions were not necessary, and re-commended a general system of levees to stretch along the lower Mississippi. The commission asserted that such a system demanded federal aid, and added that there should be a general board created on flood management, and each drainage basin needed an effective chief engineer with plenty of funding, emergency powers, and the ability to obtain rights-of-way, all to rebuild the levee system. Despite its sense of urgency, the report recommendations were never enacted. From the mid-1870s to the early 1880s, three major river conventions met to demand collective action in the Valley. The river was getting national attention, but locals waited for a substantive response.

Despite the lack of support for the proposals of the Levee Commission, Congress authorized surveys of various transportation routes in 1875. In the following year, Congress also approved construction of jetties at the South Pass of the Mississippi to create a permanent ship channel. James B. Eads, self-taught engineer and master bridge builder, successfully completed the project that provided a necessary channel for seagoing ships to the port. Eads' plan to construct jetties to remove sand bars from the mouth of the river was strongly opposed by Humphreys and other Corps officers who believed the con-struction was futile. They were proven wrong, and their reputations were damaged.

Ultimately, Congress agreed to create a permanent body to deal with the Mississippi River from a national perspective, one not dependent entirely on the work of the Army Corps, and one that united flood control interests and navigation interests. The Mississippi River Commission (MRC, with Army Corps and civilian members, and an Army engineer as president), established in 1879, denoted a major federal commitment to confronting the problems of the river. Its powers were intentionally broad but ill-defined to keep the commission from being challenged by opponents.

The commission's early history was dominated by questions concerning its functions, new surveys, preoccupation with the levee system, and a program of channel improvements through contraction of the river. The 1882 flood, which overwhelmed the levees, had a powerful effect on the MRC. The commission was given extensive power to build and repair levees, and to conduct harbor improvements in several locations. It also introduced new techniques for levee construction. Essentially, all federal works on the Mississippi were consolidated into the hands of the MRC. Local levee districts sometimes were wary of such sweeping authority but welcomed federal funding of levees.

By the early 1890s, the majority on the commission, the levee boards, Congress, and many people in the Valley came to accept "levees-only" as the answer to floods and the only flood control work attempted. This was not so much a theoretical stance but based largely on demand for an immediate and local response to flooding as opposed to something that was long-term and more wide-ranging. Levees took precedent over channel improvements. But levees failed to scour the riverbeds as some had claimed, and it became clear that levees raised flood heights. Thus in 1891–1892, the MRC turned to dredging as a way to secure all-year transport. This decision, among other things, played to economic and political interests. As historian Albert Cowdrey concluded, "Levees were simply the best established and most politically remunerative form of flood control. They were works that meant immediate protection for homes, businesses, and railroads... 'Levees only' became public policy because Congress wanted it, and, in fact, because almost everybody of influence in the Mississippi Valley wanted it" (Cowdrey, 1977, 34).

The commitment to "levees-only," however, proved to be a failure as a result of the great flood of 1927, and led to harsh criticism of MRC and substantial modification of its levees-only policy. On April 16, 1927 the Mississippi River broke through its levees at Dorena, Missouri, and three days later at New Madrid. When on April 21, the river broke through at Mounds Landing near Greenville, Mississippi, more than 100 African Americans piling sandbags by force were killed. On April 29, Caernarvon levee in Plaquemines Parish was dynamited to relieve pressure on New Orleans's levees. By the end of April, water covered more than 28,000 square miles beyond the river's banks, opened 226 crevasses (52 were major), and created about one million refugees. Somewhere between 250 and 500

Figure 14.1 Aerial photograph of flood, unidentified stretch of lower Mississippi, 1927. Family standing in windows of flooded house. Unknown author or not provided. Public domain. Wikimedia.

died in the flood. The 1927 flood was one of the great disasters in American history (Morris, 2012, 165).

Despite the devastation, the flood did not end faith in the levee system completely. Congress appropriated funds for reconstructing the levees. The Mississippi River and Tributaries Project (Project Flood) revised the levees-only approach. Under the 1928 Flood Control Act, two large floodways (a more modern term for outlet) in the Tensas and Atchafalaya basins were to be constructed, with a smaller parallel floodway at New Madrid, and a single spillway would be built to protect New Orleans. (MRC would build the Bonnet Carre Spillway in Louisiana and the Bird's Point–New Madrid floodway in Missouri to control the release of water from the river in an attempt to relieve pressure on the levees. A third spillway was added at Morganza, Louisiana.)

Other measures, such as research into "proper forestry practice," were meant to make the levee system work better, rather than replace it with something else. The levee line was to be extended north to Cape Girardeau, but outlets on the east bank of the river were not considered. Reservoirs erected on the Mississippi's tributaries were to help with flood control but also meant to deal with erosion and siltation. While these changes were substantial adjustments of "levees-only" (and not without controversy) the Corps' involvement in draining land beginning in the mid-1940s firmly entrenched the levee approach. but with modifications.

The Flood Control Act of 1928, aside from its comprehensive goal of strengthening the levee system and controlling the Mississippi, gave the federal government authority to take any lands needed for easements and rights of ways. Project Flood had brought millions of federal dollars and hundreds of workers into many small southern towns. Building levees to hold back the waters of the Mississippi began a process leading to many federal agencies operating in the area, especially those concerned with agriculture. With respect to problems caused by race, class, and poverty, very little was done, however. For example, African Americans experienced forced displacements in relief camps, there were inadequacies and inequities in Red Cross charity, and a peonage labor system grew out of the federally funded Mississippi Flood Control Project (Mizelle, 2014).

Major floods along the Mississippi River continue to this day, and levees remain a prominent tool meant to restrain them. Yet, while some extoll the value of levees to flood control, "levees-only" as a strategy has changed. Outlets designed to function like natural crevasses have been built to divert water from the Mississippi River laden in sediment into Louisiana's wetlands. Moving levees further back from the river provides more room to accommodate flood waters. As Davis said, "Now engineers are trying to reconnect the plumbing and mimic the original system" (Colten, 2000, 85).

The historic efforts, although mostly well-meaning, to force the river into a main channel walled up with levees interfered with the natural hydrology of the region. As public policy professor Richard N. L. Andrews asserted, "Levees… simply worsened flood risk: they cut rivers off from their natural overflow areas, and forced more water through narrower channels, increasing flooding downstream. The 'protected' areas behind the levees meanwhile became prime sites for economic development, increasing the damage from flooding when the levees were overtopped. By the 1930s a century of navigation improvements and levees had laid the foundation for such disasters" (Andrews, 2006, 165). Thus, levees may result in more severe floods (upstream, downstream, or on the opposite bank) which leads to more or higher levees, creating—as some suggest—a "hydrologic spiral."

Small, regular floods covering riverside floodplains—actually part of the river—were important for the health of a river itself. The silt carried over the banks produced rich soil for farming, and water spread over the floodplains created seasonal wetlands important for fish and bird habitats. Confinement of a river into a single channel pushed sediment accumulated on alluvial lands into the main channel and eventually settled along the riverbed, thus raising the level of flooding. Dams and other structures on the river blocked the flow of sediment and nutrients needed downstream. Shifts in the river due to levees and dams change the river's sediment flux to the gulf coast, alter features of the delta, increase the loss of wetlands, and can accelerate pollution from runoff—essentially altering the Louisiana coastal landscape.

Flood control on the Mississippi River placed the natural flow of the river against attempts at technical solutions to the problem. At the heart of the story was not the success or failure of technology to "harness" the river, but to what end? Human settlement and economic development along the river were long-term goals, and people wanted to accomplish them on their own terms. Without taking into consideration the environmental implications, success or failure of levees were and are measured against human wants.

Christopher Morris was right when he asserted, "The problem was that the river was predictable. Solving problems tended to cause other problems" (Morris, 2012, 97).

Many years earlier, Mark Twain in his memoir *Life on the Mississippi* (1883) concurred:

> One who knows the Mississippi will promptly aver—not aloud, but to himself—that ten thousand River Commissions, with the mines of the world at their back, cannot tame that lawless stream, cannot curb it or confine it, cannot say to it, Go here, or Go there, and make it obey; cannot save a shore which it has sentenced; cannot bar its path with an obstruction which it will not tear down, dance over, and laugh at.

Having confidence in one flood control measure over another did not necessarily mean success. After all, to what degree was flood control of any kind the only answer to the proper flow of a river? Who or what will be protected or who or what will not?

Notes and Further Reading

Andrews, Richard N.L. (2006) *Managing the Environment, Managing Ourselves: A History of American Environmental Policy* (New Haven, CT: Yale University Press).

Barry, John M. (1998) *Rising Tide: The Great Mississippi Flood of 1927 and How it Changed America* (New York: Simon & Schuster).

Billington, David P., Donald C. Jackson and Martin V. Melosi (2005) *The History of Large Federal Dams: Planning, Design, and Construction in the Era of Big Dams* (Denver, CO: U.S. Department of Interior, Bureau of Reclamation).

Camillo, Charles A. and Matthew T. Pearcy (2004) *Upon Their Shoulders: A History of the Mississippi River Commission from its Inception through the Advent of the Modern Mississippi River and Tributaries Project* (Vicksburg, MS: Mississippi River Commission).

Clay, Floyd M. (1983) *History of Navigation on the Lower Mississippi* (Washington, DC: Army Engineer Water Resources Support Center, Institute for Water Resources).

Colten, Craig E. (2000) *Transforming New Orleans and Its Environs: Centuries of Change* (Pittsburgh, PA: University of Pittsburgh Press).

Cowdrey, Albert E. (1977) *Land's End: A History of the New Orleans District, U.S. Army Corps of Engineers, and Its Lifelong Battle with the Lower Mississippi and Other Rivers Wending*

Their Way to the Sea (New Orleans, LA: U.S. Army Corps of Engineers, New Orleans District).

Mizelle, Richard M., Jr. (2014) *Backwater Blues: The Mississippi Flood of 1927 in the African American Imagination* (Minneapolis, MN: University of Minnesota Press).

Morris, Christopher (2012) *The Big Muddy: An Environmental History of the Mississippi and Its People from Hernando De Soto to Hurricane Katrina* (New York: Oxford University Press).

Reuss, Martin (1998) *Designing the Bayous: The Control of Water in the Atchafalaya Basin, 1800–1995* (Alexandria, VA: Office of History, U.S. Army Corps of Engineers).

15 Salmon, Hydropower, and the Fraser River

The history of the Fraser River in British Columbia (BC) is a story of contested resources but one with a twist. It is a story mixing the fate of the salmon population, the push for hydroelectric power, and the complication of several transboundary issues. Geographer Matthew D. Evenden stated, "Overlapping resource demands made the Fraser River a contested site of development politics. By the late 1940s, the fish vs. power debate had begun" (Evenden, 2004, 15).

Water is, among other things, a food delivery medium. For humans and animals alike, food is the primary source of energy as historian Richard White has emphasized in *The Organic Machine* (1995). Salmon have been a prized resource in the Pacific Northwest for generations. Canada's Pacific coast is vast (7,180 miles) and well sheltered, and the fishing grounds extensive. Salmon ruled the day, but herring and halibut also have been valuable catches. The five commercially exploited species of Pacific Salmon—sockeye, pink, chum, Coho, and Chinook—are all anadromous. They spawn in fresh water of the BC river systems and journey to the ocean where they may experience extensive migrations. When mature, they return to their home stream to spawn and thus complete their life cycle.

In the early twentieth century, power interests sited dams on tributaries of the Fraser for mining, forestry, and hydroelectric purposes. Yet, main-stem projects never materialized, and salmon flourished there. The Fraser remained undammed. Among Canada's largest rivers, only the Mackenzie shares that good fortune. By contrast, the Columbia River to the south of the Fraser sported an extensive network of hydroelectric dams blamed for the decline of salmon. How did the Fraser avoid the fate of the Columbia, and how did the salmon population fare as a result?

Compared to the Atlantic and Great Lakes fisheries, commercial fishing along British Columbia's coast is relatively new. Salmon proved to be important to the trading posts established near the headwaters and mouths of the rivers there. Like elsewhere in Canada, First Nations people practiced fishing for thousands of years before white settlers arrived. The rich salmon rivers such as the Fraser and the Skeena were particularly important. While whites controlled the commercial industry, native peoples were among the first workers

DOI: 10.4324/9781003041627-21

in the fishery in the nineteenth century. Local indigenous families fished and worked at the canneries in the summer—women on the canning lines and men fishing from small boats. Other groups joined the workforce, including not only Europeans but also Chinese who worked in the canneries and Japanese who fished along the coast. However, whites were loath to allow Asians to compete for their jobs on the water in many cases. Canners of the Fraser employed Japanese immigrants in their canneries and used them as fishers to suppress the wages of white workers. (Taylor, 1999, 137)

Commercial salmon fishing grew after the gold rush in British Columbia with the first canning activity probably at Alberni in 1860. Not until production techniques were improved and markets expanded did the industry take off in the late 1860s. Fishery earnings in BC ultimately outstripped those in eastern Canada.

The completion of the Canadian Pacific Railway was central to the success of the salmon industry. Prior to 1890 limits on available capital put several plants into bankruptcy, but the economic stability of the business improved throughout the 1890s. In 1902, the British Columbia Packers Association acquired 41 canneries while closing 19.

The salmon industry experienced many ups and downs in subsequent years. Regulations were lax until serious decline in the fish stocks forced a change. Periods of squabbling and cooperation between Canada and the United States took place from time to time over shares of the respective catch. Like other fisheries in Canada, the Pacific fishery faced important interludes of decline due to overfishing, destruction of fish habitats, and climate change. By the 1990s, salmon fishing was in crisis as fish populations fell to historic lows.

Spanish explorers found their way to the Fraser River in 1792, and Sir Alexander Mackenzie followed in August 1793. However, the Fraser River was named after Simon Fraser, who led a North West Company expedition almost to the mouth of the river in 1808. The Fraser is sometimes called the Salmon River, and in the Halqemeylem language, the *Sto:lo (or Staulo) and in the Dakelh language, the Lhtakoh.*

Over the years, the Fraser River has been the most productive salmon river in the world, that is, the greatest single-river producer of Pacific salmon. Filmmaker Richard C. Bocking observed:

> ... the Fraser River is the central fact of life in British Columbia. Ten thousand years of aboriginal tradition are written in its flow; two centuries of high drama floated the Fraser's currents as more recent arrivals searched, dug, hunted, hammered, plowed, chopped and built new lives for themselves along its banks. Pulsing through the heart of British Columbia, the Fraser's life-sustaining power has endowed the province with a unique rhythm. There may be no place in the world where so much diversity and beauty is woven into a single tapestry by the flow of a river.
>
> (Bocking, 1997, 2–3)

The romance of this passage, however, cannot camouflage the constant vigilance that was necessary to preserve and protect salmon as the most important resource of the Fraser. In one of the first recorded references to salmon on the upper Fraser, MacKenzie observed during his expedition: "The salmon were now driving up the current in such large schools, that water seemed, as it were, to be covered with the fins of them" (Roos, 1991, 5).

The Fraser is the longest river in BC, and one of the longest montane rivers in the world. It originates at Fraser Pass at Blackrock Mountain on the western side of the Rockies and then flows for 854 miles to the Strait of Georgia at Vancouver. The river's course is initially northwestern and then bends southward. Its catchment area covers one-quarter of the area of British Columbia (about 92,000 square miles), and about 70 percent of the region drained is more than 3,000-feet high. The Thompson River (about 145 miles from the mouth of the Fraser) is the most important of the numerous tributaries.

First Nations people lived in the Fraser River Basin for more than 10,000 years, relying on the river for resources, transportation, and trading. Salmon were of central importance to their diet and to their cultural traditions. Sockeye salmon also were the focus of the economy and lifestyle of early white settlers. In the late 1820s, the Hudson's Bay Company began buying salmon from the tribes and exporting salted salmon to the Hawaiian Islands and Asia after the building of the Fraser's Fort Langley. For the most part, the area remained a remote fur-trading area through the early nineteenth century.

Miners from California, eastern North America, and Asia flocked to the Fraser Canyon beginning in 1858 with the first big gold strike on March 23. By June, 10,000 people moved up the river to Hope and beyond. On June 7, the *Surprise*, the first steamer to reach Hope, docked with another 15,000. About 30,000 people in all arrived in 1858. A number of makeshift rafts made their way to Yale, although hazardous waters led to several deaths. With most of the land claimed around Yale, miners pressed northward through the perilous Hells Gate (a narrow gorge in the canyon 130 miles from the mouth of the Fraser) and encounters with local Indians resulted in the burning of their village and the killing of many of their people.

The gold frenzy led wealth seekers to press on to the confluence of the Thompson River. By the end of the year, the gold rush stalled because so much land was declared for claims, the river water was running high, and some areas above Yale were too difficult to access. But the rush was on again the next year. The wildness of parts of the Fraser was being transformed by miners, merchants, and settlers.

In response to the influx of people, the British created the colony of British Columbia (formerly Simon Fraser named the mainland district New Caledonia), which became a province of Canada in 1871. The Fraser Basin existed within territory that the United States ceded to the British in 1846 through the Treaty of Washington. New settlement continued along with the influx of missionaries from several denominations, but also with the resettlement of native peoples.

As settlers established communities and towns in BC, the initial mode of transportation from the mouth of the Fraser to Yale was the paddle wheeler. In the winter of 1861–1862, Royal Engineers began work on "the wagon road" which would extend by 1864 from Yale to the Cariboo goldfields (about 385-mile long). To better connect the country of Canada east to west, the Canadian Pacific Railway established a route in BC through the Rocky Mountains, over the interior plateau, along the Fraser and Thompson rivers, and to Vancouver. It was an arduous task, fraught with numerous physical obstacles and tensions between white and Chinese workers.

By spring 1884, the railway occupied the Fraser Canyon on the left bank of the river and continued to push on until completion in 1885. Construction of a railroad line by the Canadian Northern Railway along the left bank began in 1911 and was completed to the east at Ashcroft in 1915. It was later known as the Canadian National. The primary exports from the basin via the railroads were forest products and fish, the markets for the goods being widened by the cheap and accessible transportation.

As we passed up the Fraser River, to fish the Thompson, we saw, from the train, the Indians catching salmon in the great canyon above Yale.

Figure 15.1 "As we passed up the Fraser River, to fish the Thompson, we saw, from the train, the Indians catching salmon in the great canyon above Yale." John Pease Babcock, 1908. Public domain. Wikimedia.

New railroad construction in 1911–1912—especially dumping of large amounts of rock into the river—prompted circumstances that led to a disastrous impact on salmon runs in the Fraser. In addition, there was a large granite rockslide at Hells Gate in 1914 that narrowed the channel below prevented sockeye salmon from reaching historic spawning grounds. Efforts to remove some of the debris were completed by the following year.

European settlement in British Columbia disrupted and decimated First Nations communities through disease and occupation and had a direct impact on salmon stock. Mining, farming, logging, and industrial activity began to alter the aquatic environment and indigenous fisheries. Canning in particular pulled millions of salmon from their habitat creating a dynamic but threatening economic force. In 1866, the first salmon cannery on Columbia River, the Hapgood, Hume and Company, opened for business. Also, in that year the first cannery on the Fraser was built at New Westminster. By 1883, more than 50 canneries operated in the Columbia River basin, and more than 100 canneries in BC. Between 1870 and 1890, the canning industry was intensely competitive. Sockeye runs in the Fraser were enormous in 1877, 1881, and the late 1880s.

At the time the industry supported about 2,600 commercial fishers, who worked to supply salmon for the canneries. Believing salmon supplies were limitless, industry practices proved wasteful—large portions of the fish were discarded, by-catch of other species tossed aside. Depletion in various locations after the turn of century did begin to erode the notion of an endless supply of fish.

At first, salmon were pulled from the river by nets and other rudimentary technology, but ultimately the fish were being intercepted before entering the river after their long life-cycle travel. The commercial fleet largely was powered by sail, but steam-powered freighters transported the salmon from the canneries to rail heads where the product was shipped to markets. By the 1890s, the American catch of Canadian-bred sockeye began to cause concern in British Columbia. As several experts made clear, salmon do not respect international borders.

Economic development and subsequent settlement through farming, mining, logging, the impact of railroad development, the rise of the canning industry, concerns about overfishing, and competition among nations for the salmon catch contributed to the first efforts to protect the valuable aquatic resource. The initial formal discussion between the United States and Canada concerning the sockeye fishery problems related to the Fraser River system took place in Washington, D.C., in October 1892. The findings that grew out of the meetings and subsequent report, however, stated that there was no serious problem in the Fraser River District.

During the late-nineteenth and early-twentieth centuries concern about overfishing, cannery production, and decline in salmon stocks (sockeye and pink)—especially after the rockslide at Hells Gate in 1913—continued. Hatcheries were established to raise fish in captivity, and various plans were developed to keep wild populations up each season. How the dwindling sockeye catch was distributed between the two countries also became an important issue. From 1891 to 1900, Canada took about 70 percent of the total sockeye catch. During the next decade, the United States was taking 56 percent.

Various fishing methods were discussed, but a growing concern was the interception by Alaskan fishers of salmon bound to and from BC, Oregon, and Washington. The life cycle of anadromous salmon which took them across national borders was at the heart of the problem. Canadian fishers were capturing Coho, Chinook, and other species headed for rivers in Washington and Oregon. Fishers in northern British Columbia intercepted salmon returning to Alaska. American fishers caught salmon as they traveled through the Strait of Juan de Fuca and San Juan Islands toward the Fraser River. This technique to taking salmon substantially reduced the number of fish returning to the Fraser and other rivers of origin.

Various attempts to secure a mutual agreement to deal with control of the international waters with respect to fishing had its ups and downs through most of the 1920s, complicated by the nature of the salmon runs themselves. Where transboundary diplomacy failed some Canadian and American fishery agencies at least tried to cooperate. By 1928, sockeye runs had diminished, and many canneries were shut down. In response to declining salmon populations in the Fraser River, Canada and the United States negotiated the Pacific Salmon Convention (Sockeye Convention) in May 1930 and established the International Pacific Fisheries Commission (1937–1985) to regulate harvests within "the Convention waters."

Canada agreed to allow the United States to harvest 50 percent of Fraser River sockeye (and pink) stocks in Convention waters in exchange for financial and technical assistance reopening Hells Gate to salmon migration through fishways. The agreement also called for rehabilitating the Puget Sound-Fraser River sockeye salmon through fishways, restocking hatchery fish, and restricting fishing.

While significant, the 1937 convention did not resolve transboundary issues concerning fishing, nor, in and of itself, the complicated problems related to preserving the salmon. For example, historian Joseph E. Taylor III asserted, the agreement did not stop fishers from both countries "leapfrogging" one another by going out further in the ocean to capture the fish. Pushing fisheries into the ocean complicated fish management and even altered the evolution of the fish (Taylor, 1999, 201; Taylor, 2002, 175).

Transboundary disputes were not the only source of worry about sustaining the salmon runs. The rivers in the Pacific Northwest like elsewhere were subjected to contested forces none more problematic than damming and otherwise obstructing the river's flow for a variety of purposes, but ultimately for the production of hydroelectric power. Surprisingly, the Fraser escaped this fate along its main stem, but the pressure was constant over the years to bring it in line with other productive salmon rivers such as the Columbia.

All the rivers of the region experienced some type of issue with human-made obstructions going back to the nineteenth century. Beginning in the early 1870s, loggers constructed makeshift splash dams on the lower Columbia and its tributaries (the average lasting 20 years) to float logs downstream. Having very few fishways, these dams blocked passage of the salmon and reduced spawning areas. Taylor added, "Because loggers regulated the stream to facilitate transportation, flows fluctuated wildly. When they closed the dam gates, streams dried to a trickle and then were inundated by scheduled flash floods. The sudden rush of water and logs scoured spawning beds and re-channeled streams, harming any salmon, eggs, and juveniles caught in the torrent." Plus, sawmills dumped tons of sawdust into streams and bays causing incredible pollution (Taylor, 1999, 56–57). By 1910, there were 56 splash dams on the lower Columbia.

In the twentieth century, several of Washington State's rivers were dammed to provide electrical power for municipal water supplies, for farming irrigation, and for the creation of lakes for recreation. On the Columbia River—which at one time was the most productive salmon river in the world—16 main-stem dams were constructed, and on the lower Snake River (Columbia's largest tributary) another five. The gigantic Grand Coulee Dam (completed in 1941) blocked access to hundreds of miles of spawning ground on the upper Columbia and devastated tribal fishing. While other dams had fish ladders (several pools built like steps to enable fish to bypass a dam), Grand Coulee had none.

In many cases, rivers were channelized, straightened, or hardened to improve navigation or prevent flooding. The Bonneville Dam (completed in 1937) on the lower Columbia flooded a large amount of habitat and was a noteworthy barrier to upstream passage of adult salmon and downstream migration of juveniles despite the construction of fish ladders. By 1960, Chinook salmon runs in the area were less than 10 percent of what they had been a century before. On the Snake River (where four dams were built between 1962 and 1972), the major stock of Chinook salmon spawning there was placed on the Endangered Species list.

Unlike the Columbia, the Fraser was spared hydropower development on its main stem but not for the lack of trying by some. Economic growth in BC during and after World War II stimulated new demand for electrical power. As Evenden suggested, "In response, BC Electric launched a major building campaign and new developers entered the scene in the hopes of harnessing British Columbia's waterpower wealth…Overlapping resource demands made the Fraser River a contested site of development politics. By the late 1940s, the fish vs. power debate had begun" (Evenden, 2004, 15).

The Aluminum Company of Canada, in the first major postwar dispute, gained rights to develop electricity on the Nechako River (an upper-basin tributary). Several other plans for dams were proposed in the late 1940s and 1950s, but none were approved for the main stem. In the late 1940s, as pressure increased in BC to begin major dam development, salmon runs rebounded from previous lows and thus cut off some projects.

The most ambitious project involved Moran Canyon in the 1950s envisaged to be about 285-feet high and with a reservoir of about 167 miles with the largest power-generation capacity in North America. After a protracted battle, the project finally was defeated in 1972, which influenced other cancellations. There were several reasons for the termination of the project including the salmon runs were important to both Canada and the United States; scientific analysis of possible mitigation schemes questioned the effectiveness of fish ladders and other practices; opposition to the dam was well-organized; and alternative sites were available other than the Fraser (Ferguson and Healey, 2009, 4).

Development politics differed on the Columbia versus the Fraser in the 1950s. At the turn of the century, developers of hydroelectricity in BC had sought rivers and lakes close to the market in Vancouver, thus the relative remoteness, expense, and impracticality of tapping the Fraser was significant (added to this was the lack of state financial support). With new demand for electricity in the World War II era and dams popping up along Fraser's tributaries, the question of "How long will it be before the main stem is tapped?" became more pressing.

Local anti-dam forces were not influential enough to stop the momentum in the 1950s. More significant was new construction of sites on the Columbia and Peace Rivers, and their relationship to rising transnational political interests. Evenden argued: "Those alternatives on the Peace and Columbia came about in part because of the political pressure to conserve the Fraser as salmon spawning habitat. The Peace and Columbia development agenda thus operated as both a political outcome of, and a solution to, the fish vs. power conflict on the Fraser" (Evenden, 2004, 228).

Simply put, the United States had a very strong interest in the Fraser River fishery. Damming the Fraser would severely reduce salmon stock important to American fishing interests and weakened the appeal of competing hydroelectric projects especially on the Columbia. For the Canadians, development on the Peace River under BC Hydro—while leaving the main stem of the Fraser to produce salmon—answered the need to spur economic development through hydroelectric power production.

Whatever solutions seemed to be reached by the 1960s between the United States and Canada did not lead to resolution of problems related to salmon runs, interception, and preservation of a valuable resource. As Joseph Taylor argued:

> Overfishing did occur, but it happened because of concurrent social, economic, cultural, and environmental changes. Dams did provide the coup de grace, but they have been reducing salmon habitat since the 1840s. Thus the fisheries took too many fish not simply because there were too many nets in the water—that is a tautology—but because many more activities began to stress runs with the arrival of Euro-Americans.
>
> (Taylor, 1999, 39)

The rivers most often were treated separately, not part of a comprehensive plan. In 1964, the United States and Canada signed an agreement to manage the Columbia River Basin. In it, the Americans could utilize a large water-storage area in Canada for power demands and also for flood control. Canada, in turn, would receive a share of power produced and payment for its storage accession. In the 1970s, the politics of native land claims and access to resources including the fisheries became extremely important and could no longer be dismissed in any treaty negotiations.

In 1985, the U.S.–Canada Salmon Treaty was signed to help rebuild salmon runs by limiting offshore catches of some species and—again—to limit interceptions. By 1994, relationship between the two countries over salmon was strained. From 1985 to 1993, American interceptions of Canadian salmon (in part taking advantage of the fact that there were no dams on the main stem of the Fraser) increased from 6 million per year to 9 million (Mighetto and Ebel, 1994, 184–185).

The period from roughly 1992 (when the first Pacific Salmon Treaty expired) to 1999 has been called the Pacific Salmon War. It occurred because negotiations to reach agreement on a comprehensive, coast-wide fisheries arrangement went nowhere. In 1999, there was an emendation of the Pacific Salmon Treaty into the Pacific Salmon Agreement, and its renewal took place every ten years to the present with efforts at reviewing stock status and the effectiveness of the regulations.

Yet the free-flowing Fraser's special role in the history of salmon fishing in the Pacific Northwest did not mean its fish population continued to prosper while all declines took place in the dammed rivers. Sockeye returns on the Fraser in 2019 may be the lowest on record—about 600,000 as compared to the expected 5 million. Several factors account for the variations including interceptions, a variety of natural occurrences (including climate events like heavy rains and snowmelt), topography, landslide into the river, fishing regulations, and more.

The common assumption that the restrictive passage of salmon past dams was the primary culprit in reducing fish stocks has been questioned in some studies concerning the Columbia and Snake Rivers. One issue raised is the degree to which ocean migration (especially with climate change) might play a role in survival. Studies suggest that survival during the downstream migration of some stocks has been greater on the Columbia and Snake Rivers than on the Fraser. The fate of wild salmon on all the rivers, however, has been altered significantly over the centuries because of human action.

Especially since the arrival of white settlers to the Pacific Northwest, salmon were not merely a source of food and energy but a commodity to be exploited in the marketplace. That exploitation not only had a massive impact on the salmon population itself but also on the rivers and migration paths they occupied. That the Fraser River avoided the most egregious forms of development was an incongruity, but also a political tradeoff between competing nations—although it did not wholly protect the massive salmon runs of the past. The debates and disagreements pitted contention over two commodities—fish and electrical power. One form of energy versus another.

Notes and Further Reading

Bocking, Richard (1997) *Mighty River: A Portrait of the Fraser* (Vancouver: Douglas & McIntyre).

Evenden, Matthew D. (2004) *Fish versus Power: An Environmental History of the Fraser River* (New York: Cambridge University Press).

Ferguson, John W. and Michael Healey (2009) "Hydropower in the Fraser and Columbia Rivers: A Contrast in Approaches to Fisheries Protection," *Catch & Culture* 15 (May): 1–7.

Gough, Joseph and Erin James-Abra (2015) "History of Commercial Fisheries," *The Canadian Encyclopedia*, July 23, online, https://thecanadianencyclopedia.ca/en/article/history-of-commercial-fisheries.

Mighetto, Lisa and Wesley J. Ebel (1994) *Saving the Salmon: A History of the U.S. Army Corps of Engineers' Efforts to Protect Anadromous Fish on the Columbia and Snake Rivers* (Seattle, WA: Historical Research Associates).

Roos, John F. (1991) *Restoring Fraser River Salmon: A History of the International Pacific Salmon Fisheries Commission, 1937–1985* (Vancouver: Pacific Salmon Commission).

Taylor III, Joseph E. (1999) *Making Salmon: An Environmental History of the Northwest Fisheries Crisis* (Seattle, WA: University of Washington Press).

Taylor III, Joseph E. (2002) "The Historical Roots of the Canadian-American Salmon Wars," in John M. Finlay and Kenneth S. Coates, eds., *Parallel Destinies: Canadians, Americans, and the Western Border* (Seattle, WA: University of Washington Press), 155–180.

White, Richard (1995) *The Organic Machine: The Remaking of the Columbia River* (New York: Hill & Wang).

Part VI

The Postwar Years

In the years after World War II, the United States became the world's leading economic and military power. But that fact alone did not shape water issues for North America. Seminal events such as the Civil Rights movement, the consumer boom, the Cold War, and concerns about weather-produced disasters were extremely important. Significantly, the water episodes in this section augments some recurring themes found elsewhere in the book.

Linking civil rights and swimming appears curious on the surface, but there were few activities in the workplace, in the community at large, and at recreational facilities that were free of racism. In the postwar years, the rise of the Civil Rights movement drew attention to racial hostility and racial inequality that was chronic in the United States and elsewhere. The end of slavery was insufficient to rid us of prejudice, and post-Civil War segregation was the greatest reminder of how we failed to deal with unfairness and injustice directed at people of color. The case of race and swimming pools is a very useful example to consider.

Efforts to keep black people (and other people of color) out of public pools—or restricted to "colored" pools—also exposed tensions over sex and gender roles and (misguided) perceptions of health. Whites were leery of black men swimming with white girls in bathing suits, concerned about racial mixing of any kind, and convinced that black people were unclean or disease-ridden. Sometimes the exclusion came to violence instigated by whites, sometimes legal actions were taken. And just as "white flight" to the segregated suburbs offered respite from central cities and their underclasses, establishing "whites-only" private swimming clubs became another way to create white enclaves.

In a book called *Consuming Cultures: Power and Resistance* (1999), authors Jeff Hearn and Sasha Roseneil argue that "consumption is one of the basic ways in which society is structured and organized, usually unequally, sometimes incredibly so." There are clear disparities in who receives and who can afford to buy the products they want. But in the postwar years at least it appeared that many material things that people had been denied during the Great Depression and World War II, were now becoming available to the general public. In some cases, new and wondrous items entered the marketplace. One of them,

DOI: 10.4324/9781003041627-22

although seemingly mundane, was new household detergents advertised to make clothes cleaner, and ostensibly, to make life easier for housewives. This was not the first time that "labor-saving" products were claimed to change domestic life and liberate women from housework. Nor was it the first time that consumer products reinforced gender roles in the minds of the consumers.

The downside of such a consumer-product revelation, however, was to be found after its useful life. Wastewater with detergents dumped into lakes and streams was ultimately paying society back with pollution that threatened fish and made the water itself unusable. Detergent phosphates were not the first form of water pollution, but they produced results different from an array of industrial pollutants that had been with us for generations. The search for an alternative cleaning powder/liquid also brought to the fore how governments and businesses responded to the problem and the degree to which community activism could play a role in effective protest in this new age of widespread consumerism.

In the third water episode, a controversy emerged in the 1940s about treating drinking water with fluoride in order to reduce tooth decay in children. Again, universal issues are revealed in such a case. Those who opposed fluoridation might not have trusted the science behind the treatment, may have seen it as a communist plot to control people, or regarded it as overstepping of government authority. As we shall discover, there were reasons to be concerned about enthusiastic support for fluoridation in the 1940s and 1950s—the possible impulsive decision for the treatment before all the evidence was in or disregard for alternative methods of providing fluoride (such as toothpaste or mouth wash), which would make the use of fluoride a personal choice, not a mandate. In the current world of Covid-19, the debate over vaccinations, masking, and social distancing reflect many of the same issues: mistrust of science and scientists, perceived threats to individual rights, and concern about government overreach.

What makes Hurricane Hazel interesting is its unique path that led it to Toronto. We do not think about Canada when speaking about hurricanes. The surprise and lack of preparedness that occurred there was due primarily to low probability. Hurricanes don't strike Canada. They are not a problem to take seriously. A related question, however, is the insistence that weather events such as hurricanes are simply "acts of God." That is how many people view or explain them and how insurance companies often deal with them. In the essay, I quote from Ted Steinberg's book, *Acts of God: The Unnatural History of Natural Disaster in America* (2000): "Natural disasters have come to be seen as random, morally inert phenomena—chance events that lie beyond the control of human beings. In short, the emphasis has been on making nature the villain." Even in a case study about a unique event such as Hurricane Hazel attacking Toronto, Canada, is there a universal lesson to consider about understanding hurricane risk in general?

16 Racism and Civil Rights in American/Canadian Swimming Pools

Whether in rivers, lakes, the ocean, or swimming pools a refreshing dip has long been a part of warm weather. On the surface at least, recreational swimming seems to be ingrained as a perennial pastime for Americans (and Canadians)—a relatively carefree use for water for all to enjoy. And yet, as historian Victoria W. Wolcott asserted, "Swimming pools and beaches were among the most segregated and fought over public spaces in the North and South. White stereotypes of blacks as diseased and sexually threatening served as the foundation for this segregation" (Wolcott, 2019, 1).

Exclusion from using pools and beaches—which are central parts of the recreational landscape—demonstrates the persistence of racism in our country well beyond bathrooms and lunch counters, schools, and churches. Indeed, finding virulent racism in such innocent past times only helps to reinforce the malignance of the color line. In addition, largely because of such exclusionary behavior, more than half of African American children cannot swim to this day and they drown at a rate three times higher than whites.

The premier historian of the social history of swimming pools, Jeff Wiltse, argued that in the nineteenth-century North "swimming divided along several social lines, the most conspicuous being gender." "Natural-water swimming" was identified as a male activity only. Also, swimming was divided along class lines, eventually "socially acceptable among the middle and upperclasses around midcentury." Wiltse added "By the third quarter of the nineteenth century, Americans across the social spectrum plunged into water. Different classes of Americans, however, took to different waters." In rural small towns, people might share "swimming holes." In urban America, the rich could enjoy fashionable resorts or private clubs, while upper-middle and middle-class people might frequent less fancy vacationing spots along the shore, their own private clubs, or public sport and fitness facilities. Working-class and immigrant kids swam in nearby lakes and rivers or, in a few cases, public swimming pools designed for them as "large community bath tubs" (Wiltse, 2007, 13–19). In the South, Jim Crow laws after the Civil War meant racial segregation in education, voting rights, accommodations, and other areas, including swimming.

DOI: 10.4324/9781003041627-23

Municipal pools (primarily in the north) were first constructed during the last 30 years of the nineteenth century as public baths, as a way for middle-class reformers to promote cleanliness and refinement among the urban poor. In 1868, Boston opened the first municipal pool in the United States. Philadelphia was the leader in building bathing pools, operating nine—all but three in the slums. Initially, women no matter of what class did not swim in public pools. Eventually, men and women could use them on alternating days. There was no distinction between black and white swimmers, but city planners avoided building pools in predominantly African American neighborhoods. Despite the reformers' inclinations, class differences produced controversy. For example, in 1910 a proposal to build a large municipal pool in New York's Central Park received strong opposition from the middle and upper classes who feared the pool would attract "all sorts of undesirable people," meaning immigrant and working-class children.

Lacking knowledge of how disease spread by bacteria or germs, the pools themselves were regarded as cleansing agents. The early pools also did not have showers. Because of their popularity, the building and use of public pools survived even after the germ theory of disease was embraced around the turn of the century. Many new pools, however, were redesigned to focus on exercise and sports.

Wiltse concluded that at the end of the nineteenth century, two distinct pool cultures existed in the United States: "Working-class boys had created a play and pleasure-centered culture at municipal pools [they did not consider this bathing but playing]. Elite and middle-class men had developed a more serious and orderly culture at private pools that reflected the competitive and directed character of Victorian culture. Both, however, were masculine cultures that valued physical vigor and camaraderie" (Wiltse, 2007, 30).

Beginning in the 1920s, cities built thousands of leisure swimming pools (very few built especially for black residents). Before then, there were probably less than 200 municipal pools in the United States. This burst in building temporarily ended with the onset of the Great Depression but picked up again in 1934 with the construction of government-sponsored pools through New Deal programs. As societal norms were changing, men and women increasingly swam together.

Gender mixing exacerbated (or became an important cause of) de facto racial segregation at public pools, alongside exclusions in law through "Whites-Only" practices. In the minds of many whites, the presence of black men in such intimate quarters with white women posed a sexual threat. In addition, widespread prejudice built on the presumption that black people were more likely to be infected with communicable diseases—thus "dirtier"—justified segregation or exclusion at swimming pools. The same prejudices were extended to beaches and beachfront resorts.

An obvious type of de facto segregation came through intimidation and violence. A well-known incident occurred in August 1931 at the opening of Highland Park Pool (a facility with two large pools and a sandy beach) in

Pittsburgh. Thousands of kids flocked to the park pools, but every black person was asked to present a "health certificate" and then was turned away. The day after the opening, about 50 young black men were allowed to enter when Superintendent of Public Works Edward Lang was pressured to do so.

Some of the approximately 5,000 white people already at the pool shouted threats and began to beat the black youths as they entered the water. The pattern of behavior was repeated for the next two weeks; some of the black youths were hit with rocks or clubs. Police officers most typically arrested the victims for "inciting to riot." Newspaper writers and city leaders attempted to blame the violence on East End Italian "hooligans," thus viewing it as an ethnic-racial clash. "The whole trouble," according to the *Pittsburgh Courier*, the city's African American newspaper, "seems to be due to the way the Highland Park Pool is operated. It is the only city pool where men and women, girls and boys swim together." This brought "the sex question into the pool, and trouble was bound to arise between the races" (Wiltse, 2007, 128).

Over the next several years, racial violence erupted at several other public pools. Such incidences were most common at mixed-gender facilities. At the time, several states in the North had laws prohibiting racial discrimination in public places, but this did not curb the violence as officials often looked the other way. Racial segregation in the South was more explicit with "Whites Only" pools and sometimes complementary Jim Crow pools that were smaller and clearly inferior to pools used by whites. By 1938, the YMCA operated 684 pools and administered countless swimming programs. Yet, African Americans had access to very few of either. In 1928, there were only 53 "Colored" YMCA branches throughout the United States, and only 18 had swimming pools.

The famous builder of modern New York City, Robert Moses, used his power and influence to limit African Americans access to a variety of public recreation facilities including swimming pools—or at least segregated them in some fashion. Moses' biographer Robert Caro observed that "The ingenuity that Robert Moses displayed in building swimming pools was not restricted to their design." He built a pool in Colonial Park in Harlem that he determined would be the only pool that "Negroes—or Puerto Ricans, whom he classified with Negroes as 'colored people'" were going to use.

A pool in Thomas Jefferson Park, in a white district, was close to large black and Puerto Rican populations. To discourage its use by "colored people" Moses posted only white lifeguards and attendants there. As a further precaution, Caro stated, "the water was left unheated [at the Thomas Jefferson Pool], so that its temperature while not cold enough to bother white swimmers, would deter any 'colored' people who happened to enter it once from returning" (Caro, 1974, 513–514).

In some cases, short of violent interactions, African Americans might be relegated to a small indoor pool set away from a large, outdoor pool for whites. In St. Louis, where blacks represented 15 percent of the population in the

Figure 16.1 Swimming pool, Moultrie, Ga. The Tichnor Brothers Collection, Boston
Public Library. Public domain. Wikimedia.

mid–1930s, they were allocated one small indoor pool, while whites had access
to nine pools. In several cities, the pools were closed rather than admit black
swimmers. According to law professor Taunya Lovell Banks, "Many early
efforts by blacks to gain access to public swimming pools focused on securing
separate pools rather than integrating all-white pools because, under the separate but equal doctrine approved by the Supreme Court in *Plessy v. Ferguson*,
racial segregation in public places was legal" (Banks, 2014, 223). Not only
were African Americans severely limited or denied opportunities to swim, but
they also were denied opportunities to learn how to swim through lessons or
actual swimming experience.

After World War II—on paper at least—pools gradually were desegregated.
But, in practice, white people found ways to exclude black Americans from
swimming. Many southern cities simply avoided building public pools or
drained existing one and did not desegregate any facilities for many years. In
the North and West, where desegregation was making some headway, the
times that African Americans could swim in public pools was restricted, white
swimmers might abandon public pools all together, or proprietors would shut
down the pools completely. It was not uncommon in various places for staff
members to drain pools and clean them after blacks had been swimming there.

The example of Baltimore was instructive about de facto segregation at the
time. In 1953, the city operated five outdoor pools for whites and one quite
inferior one for blacks. In a 1955 decision in *Dawson v. Mayor and City Council
of Baltimore,* a federal appeals court ordered the city to end segregation at its

pools starting the next year. Official segregation was stopped, but black swimmers still did not gain access to three of the pools located within predominantly white neighborhoods, where they often were jeered if trying to enter. Especially in the South, the *Dawson* decision proved wildly unpopular (Wiltse, 2014, 374).

In Omaha's Peony Park two African American swimmers were barred from participating in an Amateur Athletic Union swim meet in August 1955. The action was illegal under Nebraska law, and Peony Park was found guilty and paid a token fine of $50. Yet, the pool remained whites–only for eight more years, and eventually Peony Park opened as a private club. Tactics for denying black swimmers access to public pools even in an age of desegregation was not simply a southern thing.

Often white attendance shrank if African Americans utilized the pools, and few new pools opened during the onset of desegregation. New York City and Washington, D.C., combined opened 19 new pools in the 1930s, but none between 1945 and 1960. Several public pools closed or simply fell into disrepair, especially in inner cities facing "white flight" to the suburbs. Recreational swimming was energized in the suburbs where private swim clubs grew in large numbers. While there were about 1,200 private swim clubs in the nation in 1950, more than 23,000 existed in 1962. Swimming lessons and swimming teams opened to white kids at these clubs were off limits to blacks who lived far away from the facilities and could not afford the high cost of membership. Hundreds of thousands of backyard pools were constructed, offering owners a measure of prestige and an easy way to determine their use (Wiltse, 2014, 375–377).

During the late 1960s, there was a renewed but short surge of public pool building. Events like the 1966 race riot in Chicago influenced some city officials to build pools as a mechanism for alleviating inner-city tensions. Four days after the Chicago riot ended, for example, Great Society programs of the Lyndon Johnson administration funded swimming pools across the country for "disadvantaged youths." Soon "anti-poverty" grants were made available for such pools in 40 cities, including Chicago, New York, Philadelphia, Atlanta, and Washington, D.C. Most of the pools were small and provided little if any leisure space, and often were too crowded to encourage swimming. Children in a New York neighborhood referred to the pools as "giant-sized urinals." Such pool building stalled in the 1970s as major cities underwent increasing financial woes. The pools deteriorated with the rest of underserved minority neighborhoods (Wiltse, 2014, 379).

African American leaders did not sit idly by while black swimmers were denied access to pools even as institutional segregationist barriers were starting to fall. In several cases, direct-action protests were bi-racial. In 1949, a St. Louis official determined that black residents had a right to use public property like pools. But a day after his action, when 50 black swimmers showed up at Fairgrounds Park, they were attacked by almost 200 white teenagers with baseball bats and large sticks. As the clashes spread, the mayor ordered the pools to be segregated again. The local NAACP sued and forced the city to desegregate the pools by race, but not by sex.

In the 1960s, black protesters staged "wade-ins" to demand equal access to a beach in Biloxi, Mississippi. In response, a riot ensued including gunfire. Not until 1968 did the federal court rule that the beach had to be open to allcomers. Civil rights protests in St. Augustine, Florida, led by Martin Luther King, Jr., and the Southern Christian Leadership Conference, included a "dive-in" at the Monson Motor Lodge where white and black activists jumped into the pool. Not only did white police officers also jump into the pool to arrest the protesters, but the motel manager dumped muriatic acid into the water near the swimmers in an attempt to burn them. In pressing for civil rights after the passage of the Civil Rights Act of 1964, the interracial group the Arkansas Council on Human Relations predicted that public swimming pools would be—along with small restaurants— "the most serious holdouts" (Kirk, 2014, 140).

In a celebrated television event on a 1969 episode of *Mister Rogers' Neighborhood*, Fred Rogers invited recurring character Officer Clemmons (who was black) to join him in soaking his feet in a wading pool. In a 2018 documentary, Francois Clemmons stated, "They didn't want black people to come and swim in their swimming pools." "My being on the program was a statement for Fred" (Chokshi, 2018).

There has been a long history of exclusion at swimming pools not only in the United States but also in Canada. The circumstances were somewhat different, but the results were the same. As stated in a recent article in *Society and Leisure*, "Despite Canada presenting itself as post-racial when it comes to Black people…it has long had to wrestle its own demons…" (Nzindukiyimana and O'Connor, 2018, 138). As in the United States, swimming had evolved into a "White activity" in the early twentieth century, but pools were mostly not segregated in Canada. People of color were excluded most often by unofficial means.

Class was a main category of exclusion in the late nineteenth century in Canada, but eventually focused on gender and race. Ultimately race became the primary category of exclusion lasting well into the twentieth century. But, as the *Society and Leisure* article explained, "In the absence of explicitly racist laws or regulations, as could be found in the neighboring US, generic and anti-discrimination principles could be and were arbitrarily interpreted" and applied (Nzindukiyimana and O'Connor, 2018, 140).

There are plenty of examples over the years of such arbitrary and random exclusion practices in Canada. This included beaches as well as swimming pools where, especially in the early years, one could find signs stating "Whites Only" or "White Gentiles Only." In 1923, the Edmonton city council ruled that black people were banned from city pools. In 1948, a pool manager at a Calgary pool refused to admit a black girl into the pool. "That's always been the rule here," he told the *Calgary Herald*. "If too many Negroes came to swim no one else would want to use the pool and we'd go out of business." He added that the rule also applied to Chinese and Japanese people. Between 1928 and 1945, Chinese people were banned from public swimming pools in Vancouver (Gajewski, Misha, 2020). In the 1960s, racial intolerance remained unresolved in Canada, and black activists—as they believed in the United States—saw swimming as a way to defy racial oppression.

The randomness and arbitrary nature of exclusion in Canada reflected, to some degree, a society with a small minority population (in the 1940s Black people represented only 0.02 percent of all Canadians) and informal racial laws. But minorities—including indigenous people—were regularly pushed to the margins of society for the convenience of whites. And if the incidences of exclusion were less frequent and the level of violence much less intense (or non-existent), "The manifestation of racial prejudice in swimming spaces was an extension of discrimination encountered in other spaces" (Nzindukiyimana and O'Connor, 2018, 157).

Turning back to the United States, discrimination in the use of public swimming pools—as we have seen—did not end with desegregation, in the same way that many other forms of discrimination did not end with important legislation such as the Civil Rights Act of 1964. De facto segregation and hostility to minority rights festered in many places up to the present. Prior to 1971, the Supreme Court did not rule on the legitimacy of pool closures as a way to exclude African Americans. In *Palmer v. Thompson*, however, it directly addressed whether the closing of a Jackson, Mississippi municipal pools "in whole or in part to avoid desegregation…was a denial of equal protection of the laws." In a 5-4 decision it agreed with the lower courts' rulings that Jackson could discontinue the facilities rather than desegregate them.

Such a decision encouraged the decline of public pools and the increase in private swim clubs. As a 2018 article stated, "The legacy of the civil rights era litigation championed in [*Brown v. the Board of Education, Topeka*] was not extended to public pools in *Palmer*…Because the Supreme Court did not view pools as essential public facilities, the problem still persists today" (Banks, 2014, 230–234; Waller and Bemiller, 2018, 10).

In some cases, economic conditions reduced swimming opportunities as well when pool closures increased. After the recession of 2008, a wave of pool closures across the country was significant, especially impacting poor and working-class children. Also recurring cases of discrimination against African Americans (and Latinx people as well) at swimming pools still periodically make the press. In 2009, a suburban swimming club in northeast Philadelphia revoked a contract where a youth camp was to be held (including black and Latinx kids), stating that the pool was unsafe with so many kids participating. A state commission sided with the camp, and the club agreed to a settlement.

In 2014, a story surfaced concerning La Vacherie Swimming Club in Louisiana, when one of the members of the River Parish Swim League was kicked out of the league. About 30 percent of that team was black. La Vacherie was rigidly segregated, requiring a membership be purchased by someone who was selling theirs. White applicants moved quickly off the waitlist, while black applicants never came off the list. In 2015 in a wealthy subdivision in Dallas, police targeted black teenagers attending a pool party. To this day, newspapers and the Internet regularly report on incidences of harassment of those "swimming while black."

There are, of course, feel-good stories like that of Simone Manuel, who at 20 became the first African American woman to win a gold medal in an individual event (100-yard freestyle) in Olympic swimming history at the 2016 Rio Games. A few days after Manuel's feat, goalkeeper Ashleigh Johnson on Team USA became the first black woman from the United States to win a gold medal in water polo. No matter that for centuries Africans had been well-known for aquatic achievements; even in the world of Simone Manuel and Ashleigh Johnson racism undercut the opportunity for black people to learn how to swim, to swim recreationally, or to compete in water sports.

Accidental drowning, in recent years, is the leading cause of death for black youths. Indigenous people in Canada also represented a disproportionate number of people who drowned (between 2009 and 2014, about 10 percent). In an incident in Louisiana in 2010 sometimes repeated elsewhere, six black teenagers who could not swim drowned in a river trying to save a friend while their horrified parents—also who could not swim—looked on. Trying to rectify the swimming dilemma, groups such as "Black People Will Swim" have sprung up in various communities to end such tragedies and to provide opportunities for black children to simply enjoy the water.

What then is the connection between water, the environment, and racism in this story? Wiltse argued that there is a "uniqueness of swimming pools as physical spaces" and also that "swimming pools serve as useful barometers of social relations" (Wiltse, 2015). Possibly the intensity and endurance of racial discrimination around swimming pools have to do with their location, limited space, the intimacy of wearing swimsuits, and the unavoidable social contact in such a setting. This confluence of circumstances can bring out the worst in people, where water-related activities—be they pure recreation, part of a sporting event, or ritualized through religious and cultural practices—require a sharing not present in individualized, more isolated situations. In some respects, the communal as opposed to the individual aspect of sharing the water and this particular use of space brings out responses not so easily controlled by regulation or prohibition. Racism in this context can be particularly unsettling.

Notes and Further Reading

Banks, Taunya Lovell (2014) "Still Drowning in Segregation: Limits of Law in Post-Civil Rights America," *Law & Inequality: A Journal of Theory and Practice* 32: 215–255.

Caro, Robert A. (1974) *The Powerbroker: Robert Moses and the Fall of New York* (New York: Alfred A. Knopf).

Chokshi, Niraj (2018) "Racism at American Pools Isn't New: A Look at a Long History," *New York Times*, August 1, 2018.

Gajewski, Misha (2020) "How Systemic Racism Shaped the History of Canada's Swimming Pools," *Freshdaily*, online, https://www.freshdaily.ca/sports/2020/08/systemic-racism-history-canada-swimming-pools/.

Holland, Jearold W. (2002) *Black Recreation: A Historical Perspective* (Chicago, IL: Burnham, Inc., Pubs.).

Kirk, John A. (2014) "Going Off the Deep End: The Civil Rights Act of 1964 and the Desegregation of Little Rock's Public Swimming Pools," *Arkansas Historical Quarterly* 73: 138–163.

Nzindukiyimana, Ornella and Eilenn O'Connor (2018) "Let's (not) Meet at the Pool: A Black Canadian Social History of Swimming (1900s-1960s)," *Society and Leisure* 42: 137–164.

Rousell, Davina D. (2011) "Leadership, Power and Racism: Lifeguards' Influences on Aboriginal People's Experiences at as Northern Canadian Aquatic Facility," *Leisure Studies* 31: 409–428.

Waller, Steven N. and Jim Bemiller (2018) "Navigating Rough Waters: Public Swimming Pools, Discrimination, and the Law," *International Journal of Aquatic Research and Education* 11, 1–15.

Wiltse, Jeff (2007) *Contested Waters: A Social History of Swimming Pools in America* (Chapel Hill, NC: University of North Carolina Press).

Wiltse, Jeff (2015) "Racism at Pools is Nothing New," *Pittsburgh Post-Gazette*, June 13, 2015.

Wiltse, Jeff (2014) "The Black-White Swimming Disparity in America: A Deadly Legacy of Swimming Pool Discrimination," *Journal of Sport and Social Issues* 38: 366–389.

Wolcott, Victoria W. (2019) *The Conversation*, July 9, 2019, online, https://theconversation.com/the-forgotten-history-of-segregated-swimming-pools-and-amusement-parks-119586.

17 Detergent Phosphates in the Great Lakes

In the post-World War II consumer boom, the widespread availability of all kinds of new consumer goods—including household wonder products—was one indicator of the end of the austerity of the Great Depression and the war. With the consumer cornucopia came greater quantities of wastes from excess packaging and paper to a variety of new or more abundant chemicals and synthetic materials. In the mid-1960s and early 1970s, attention turned to a very specific effluent—the use of phosphate-based laundry detergents in the home. That something as mundane as new detergents became a battleground for debate over consumption of goods and defiling the environment is intriguing, but it was not surprising that middle-class housewives were a major force behind confronting the problem.

This essay explores the controversy surrounding the dumping of detergent phosphates into the Great Lakes. The Great Lakes have long been a major source of fresh water for Canada and the United States. They also suffered from inexorable pollution from numerous sources.

The synthetic detergents, however, posed new kinds of pollution problems and addressing them created tension among companies producing the detergents, various levels of government, and women consumers and activists.

The Great Lakes are central to the physical and economic life of both Canada and the United States. The water is essential for consumption, transportation, fishing, recreation, power, agriculture, and industry. The five Great Lakes—Superior, Michigan, Huron, Erie, and Ontario—represent the largest freshwater surface in the world and make up one-fifth of all fresh water on the globe, and over 95 percent of North America's fresh water. The lakes cover about 94,250 square miles (almost as big as Oregon) have 10,500 miles of shoreline and hold 5,439 cubic miles of water. They are interconnected by rivers, straits, and canals. The St. Lawrence Seaway (completed in 1959) links them with the Atlantic Ocean via the St. Lawrence River. With the exception of deep winter months, the passage is one of the world's busiest shipping areas. As of 2006, about 33.5 million people live in the Great Lakes—St. Lawrence River Basin—about one-tenth of the U.S. population and one-quarter of the Canadian population.

DOI: 10.4324/9781003041627-24

Several cities along the lakes were manufacturing centers in the nineteenth century, in areas such as food processing, steel production, and the automobile industry. The Great Lakes region became the center of heavy industry in the United States and to a lesser degree in Canada. Some of the world's largest concentrations of industrial capacity are in the Great Lakes region.

The Great Lakes have long been a sink for a variety of pollutants. The likely first source came from logging, particularly sawmills located at the mouth of some rivers in the northern parts of the Great Lakes Basin (especially Lakes Superior, Huron, and Ontario) and clearcutting trees along the lake shorelines. Wetlands drained for farming caused erosion and runoff. Mining waste found its way to the lakes in the northern arc of the basin. Shipping had various impacts. For example, forests were denuded to gather wood for steamboats resulting in more erosion and runoff and in air pollution from smokestacks. The use of DDT and other pesticides, PCBs, and commercial fertilizers added a variety of chemicals to the water. Mercury undermined fish populations. Human waste from sewers poured into the water from homes and businesses (Dove and Chapra, 2015, 702–703; Shear, 2006, 202–203).

Human use and transformation of the lakes' environment through pollution also included what some have called "biological pollution," which severely altered the ecological balance. Unintentionally or intentionally, the lakes have experienced more than 160 invasive, nonnative, or exotic species of zooplankton, fish, benthos, algae, and aquatic plants. Between 1819 and 1974, 34 nonnative fish species were introduced—some intentionally like the common carp from Germany, which proved destructive to other fish. Native fish populations changed dramatically in the twentieth century, first through overfishing and then through the introduction of species such as sea lamprey, which seriously reduced lake trout in Lake Huron and Lake Michigan by the late 1950s. Another nonnative fish, the alewife, was detected in Lake Ontario in 1873, likely via the Erie Canal. Zebra mussels all but replaced native mussels from the lakes. They first appeared in Lake St. Clair north of Lake Erie in 1986 (coming originally from the Black and Caspian seas), while Quagga mussels from Ukraine appeared in Lake Erie in 1989.

According to historian Terence Kehoe, beginning in the mid-1960s "Many American public officials and environmental advocates, after initially concentrating on reducing the pollution from the [Great Lakes] region's manufacturing plants, became convinced that a significant reduction in the level of Great Lakes water pollution could only be achieved through sweeping restrictions on the sale and use of a common consumer good: household laundry detergent products" (Kehoe, 1992, 21). This assertion may have been overstated, but debate and impact over synthetic—or phosphate—detergents was in many ways unique and unlike other pollution problems that plagued the Great Lakes beginning in the 1960s.

Synthetic detergents, first produced in Germany, were introduced in the United States in the 1940s, and by 1958 represented 72 percent of all detergents produced there. Although soaps caused pollution, synthetics caused more complex water-quality problems because of how they acted as a cleaner. Also, because they were used in greater quantities than soap (especially in modern washing machines that used substantial amounts of water), they found their way into sewage plants in great amounts and then into a wide variety of watercourses. When added to water, detergents remove soiling particles such as dirt and oil from the item cleaned and then attach the particles to water molecules to be whisked away in the drain.

Soap and synthetics represent the two types of detergents. Soaps are made from animal or vegetable fat and work well in soft water. Both soaps and synthetics have surfactants—surface-active agents—which were able to clean because they implanted themselves between the dirt and the item cleaned. In synthetic detergents, alkyl benzene sulfonate (ABS) was the surfactant in synthetics. It was introduced after World War II in products such as Fab (Colgate-Palmolive-Peet, 1948), Tide (Procter & Gamble, 1946), Breeze (Lever Brothers, 1947), and Surf (Lever Brothers, 1948).

Because of its design, ABS caused foaming—or sudsing—in sewage effluent. (The sudsing, equated with cleaning, was used as a positive attribute in sales promotions.) ABS also lacked biodegradability. Synthetic detergents have a "builder"—most often phosphates that soften hard water and increase cleaning effectiveness were included in heavy-duty detergents such as Tide, Fab, and Surf. In the early 1960s, surfactants amounted to 10–15 percent of synthetic detergents, while the builder made up about 60 percent (Read, 1996, 232–233).

Because of their versatility and affordability synthetic detergents soon were favored as a domestic and industrial cleaner over soap, and as they became more common in households and commercial establishments their scale of use raised serious concerns about water pollution. In the 1960s, the debate over synthetic detergents occurred in two stages: the first focused on the issue of excessive foaming and the second erupted over the role of phosphates in impairing water quality, especially in Lake Erie.

Concern rose in the Great Lakes area on the American and Canadian side in the early 1960s as unwarranted foaming inundated sewage treatment plants and began to appear in rivers and lakes used as sources of drinking water. In the late 1950s, foaming incidents elsewhere had been observed. For example, on the Ohio River near Wheeling, West Virginia. The detergent industry deflected and even denied that their products were responsible for foaming problems. An investigation conducted by sanitary engineers at the Massachusetts Institute of Technology, however, showed that surfactants used in the manufacture of the detergents, especially ABS, were the cause. Industry studies confirmed that a substantial amount of ABS from detergents passed through the plants, but they claimed that ABS posed no threat to humans in the concentrations found in surface- and groundwater. Such a response downplayed the pollution problem, instead calling it merely a question of appearance.

The industry knew that eventually it would have to replace ABS and had studied the experiences elsewhere with detergent pollution. As early as 1953, the British set up committees with industry representatives to seek solutions. In 1961, the West German government took a more aggressive position and mandated that after October 1964 only detergents that were at least 80 percent degradable could be sold. Democrats in the U.S. Congress, impressed by the German action, introduced bills banning nondegradable detergents to begin in mid-1965. This response followed on the heels of more aggressive federal water pollution legislation at the time revolving around recent amendments to the Water Pollution Control Act (1948).

Not surprisingly the American detergent industry energetically opposed the bills, not wanting to replace the relatively inexpensive ABS with another surfactant. It made the case that it could voluntarily find a solution by the end of 1965. Urged on by the legislative threat, the potential new regulations at home and actions in Europe, the industry—after testing over 750 compounds—replaced ABS with a degradable surfactant LAS (linear alkyl sulfonate), which it added to its products in June 1965 (McGucken, 1991, 8–9).

The controversy over foaming, as it turned out, was just the beginning of pressure facing the industry in producing its synthetic detergents. Since the 1960s, researchers had begun to take notice of rising phosphorous loads in the Great Lakes, especially Lake Erie. Some even claimed, inaccurately, that Lake Erie was "dead," but it was suffering nevertheless. Compounds containing phosphorus, carbon, and nitrogen enriched the aquatic environment, promoting growth of certain species of plants such as algal blooms (nuisance algae), with phosphates recognized by 1963 as the principal chemical involved in algal growth (McGucken, 2000, 33).

This led to eutrophication of their waters. Eutrophication is "a process in which lakes become progressively more enriched by an increasing supply of plant nutrients. This excessive enrichment results in the rapid growth of algae and other aquatic plants, which in turn results in oxygen depletion as the algae die off and—like any organic waste in water—generate biochemical oxygen demand. The massive algae blooms accompanying eutrophication also contribute to a general degradation of water quality..." (Kehoe, 1992, 24). Such impacts decrease biodiversity in plants and animals, reduce the watercourse's resource value in fishing, hunting, and recreation, and may cause or contribute to health problems in humans.

In general, eutrophication happens when a lake (in this case) ages over many years, and such changes are gradual as it moves from being nutrient poor (oligotrophic) to being nutrient-rich (eutrophic) (Beeton, 1963, 240). In extreme cases, eutrophication can lead to hypoxia—oxygen depletion—which threatens many species in the lake. However, cultural eutrophication (nutrients added to the water through human action) which was mainly agricultural and municipal runoff (including sewage), industrial waste, and phosphates in synthetic detergents greatly accelerated the process along the Great Lakes.

Renown environmental engineer Daniel A. Okun stated:

> Phosphates in natural water are not a contaminant or pollutant in the same sense as are heavy metals, complex organics, or radio-active materials. Phosphates may be handled without danger and may be consumed safely in virtually whatever amounts are likely to be found in water and foodstuffs. Phosphates only become a problem when they contribute to excessive *eutrophication.*

(Okun, 1972, 64)

Yet Okun asserted that "Problems associated with maintaining and improving the quality of our environment are seldom amenable to simplistic solutions." He worried "If phosphates were removed from detergents, we would be replacing an effective well-understood chemical that is completely innocuous to man with substitutes that would likely be less effective, certainly less understood, and very possibly dangerous." Rather than favoring "a uniform regulatory program," he believed that "every situation is unique...Nationwide uniform standards, while simple to administer, would in the end exact a far greater cost from society than would a system that recognizes local variations and permits regulations to be framed accordingly" (Okun, 1972, 76–77).

Imbedded in Okun's remarks were essential issues debated in the fight to remove phosphates from detergents in the late 1960s and 1970s. Was eutrophication a serious problem that outweighed the value of an effective cleaning agent for the home? Was this a local versus national problem when it came to championing regulations?

"A lack of scientific consensus on the link between detergent use and water pollution," Kehoe asserted, "made the formation of government policy on this issue extremely difficult" (Kehoe, 1992, 24). The manufacturers of soap and detergent products had a lot at stake. In 1967, more than 300,000 people were employed in the production and sale of these products. In terms of dollar value, they were by far the most important products produced by the industry. Household consumption represented 90 percent of the total volume of detergent. Procter & Gamble, Lever Brothers, and Colgate Palmolive were the "Big Three," and they belonged to the Soap and Detergent Association. It was a powerful trade group, which also included major chemical companies such as Dow and Monsanto, that supplied the detergent companies with substantial amounts of raw material (Kehoe, 1992, 26).

As in the case of the foaming controversy, the federal government—under the presidency of Lyndon Johnson—called for a cooperative approach forming the Joint Industry-Government Task Force on Eutrophication. Charles Bueltman, an officer of the Soap and Detergent Association, was named chair. However, pressure came in late 1969 from a report of the International Joint Commission (IJC) on pollution in Lakes Erie and Ontario (an organization established by the United States and Canada under the Boundary Waters Treaty of 1909). In it the commission stated that the best way to reduce eutrophication was through a reduction—and subsequent replacement—of detergent phosphates with a less harmful alternative. The research in the report estimated that phosphate detergents were responsible for 70 percent of the phosphorus in American municipal waste and 50 percent of Canadian municipal waste. The IJC recommended 1972 as the latest date to completely remove the phosphates from the detergents.

Figure 17.1 City beach sign warns against swimming in polluted Lake Erie. To reduce pollution, lifeguards poured chlorine into the water (June 1973). Frank J. Aleksandrowicz, photographer. Public domain. Wikimedia.

The response of the industry followed familiar corporate lines in other businesses: There was scientific uncertainty about the issue of lake eutrophication and its causes (less than asserted), and the possible harmful effects of phosphate substitutes. Even if phosphates were a problem, what about the large amounts entering the lakes through runoff from fertilized land? As the debate went on, the Richard Nixon administration followed the federal government's previous stance by placing trust in the detergent industry to find a solution. The industry, following a contradictory path, continued to defend the value of phosphates in its products but declared—in a somewhat contradictory way—that it was working diligently on finding a substitute.

With stasis on the issue at the national level, the debate turned local. In winter 1970, the IJC held six public hearings in Canada and the United States on its pollution report about the lower Great Lakes with the phosphate issue at the center of discussion. Increased media coverage of the eutrophication problem linked to synthetic detergents sparked growing grassroots action.

Housewives had been the center of marketing activity by detergent manufacturers since their product went into stores. Domesticity still was equated with specified gender roles. It was not surprising that a group of women in Buffalo founded Housewives to End Pollution in spring 1970 in response to the IJC findings and sought to put pressure on local markets and detergent manufacturers. According to historian Annette Mary Scherber in her study of the phosphate debate in Indiana, "In the midst of this conflict, women's private laundry practices as well as their opinions on how to abate water pollution became increasingly sought after by the media, legislators, and environmentalists." Her study considered how "gender framed the conflict about how to abate phosphate-based detergent pollution...and the popularization of ecological consumption, or the conscious practice of buying products that cause the least impact on the environment, in North America during the 1970s" (Scherber, 2018, 3).

The debate over phosphates did indeed revolve around consumer practices, especially those of women regarded as the central players in domestic housekeeping. Prevailing gender roles, as much as environmental practices at the time, targeted women as among those most aggrieved by a dependence on a key polluting agent like phosphates in the Great Lakes. But women were also crucial, as Scherber argued, as influential in enacting key water-quality policy "by distributing lists of phosphate content of detergents in grocery stores, lobbying government officials, writing letters to newspaper editors and politicians, speaking out at local hearings, and forming activist groups dedicated to pollution abatement" (Scherber, 2018, 175).

Women activism, of course, was not limited to this one controversy. Going back at least to the nineteenth century in the United States women had been central to environmental reform efforts over a range of sanitation issues (often referred to as "Municipal Housekeeping") and health questions. Into the twentieth century they led several movements including protests over toxic materials at places such as Love Canal in New York in the 1970s. Not all of the protests had to be gender specific, however.

Not surprisingly, the first legislation to regulate detergent phosphates in the United States occurred on the municipal level. In October 1970, Chicago became the first American city to limit and then ban the sale of phosphate detergents in the city. Akron and Detroit soon followed, and then others. In April 1971, Indiana was the first state to pass a phosphate control law (Kehoe, 1992, 32). On the federal level, Canada moved toward banning phosphates nationwide but in stages beginning in 1970–1972, while American governmental agencies debated whether NTA (nitrilotriacetic acid) was a suitable replacement for phosphates. Studies suggested that NTA might break down into toxic substances in water.

The Big Three had invested millions of dollars in research on detergents containing NTA. Given the brewing controversy, it shelved the new products and sought to develop an alternative for nonphosphate detergents. In the meantime, the major producers publicly criticized the cleaning effectiveness and safety of several nonphosphate detergents being produced by smaller companies (such as metasilicates and carbonates). The government's failure to support NTA, rising concern among the public over nonphosphate detergents, and the Big Three's media campaigns blunted an effort for a nationwide ban on phosphates. In fact, on the federal and state level, calls to use phosphate over nonphosphate detergents set off a most confusing dispute of its own, especially among those entities that banned phosphates and those that did not.

Unlike Canada, the United States did not enact a law banning phosphate detergents, but a nationwide voluntary ban came about in 1994. By 2010, 17 states banned phosphates from dishwasher detergents. They are Illinois, Indiana, Maryland, Massachusetts, Michigan, Minnesota, Montana, New Hampshire, New York, Ohio, Oregon, Pennsylvania, Utah, Vermont, Virginia, Washington, and Wisconsin.

In recent years, claims have varied as to what extent regulating phosphates—and other sources of pollution—have led to cleaner Great Lakes. "The system of cooperative pragmatism that had developed over decades in the Great Lakes Basin came apart during the 1960s," stated Kehoe, "to be replaced by a regulatory regime that differed in fundamental ways" (Kehoe, 1997, 7).

One might have thought the advent of the environmental movement in the 1960s and greater attention to water-quality issues meant more stringent and more cohesive regulations. Be it the nature of this consumer-driven controversy over phosphates or the jumble of contesting parties involved—women consumers, various layers of government (Canadian and American), IJC, scientists and engineers, and grassroots organizations—a clear resolution did not and probably could not have occurred. The debate over phosphate detergents and the Great Lakes unveiled several unique qualities, but it also revealed the confounding problems associated with building consensus over environmental pollution. The contesting forces make it difficult to unmuddy the waters.

Notes and Further Reading

Beeton, Alfred M. (1963) "Eutrophication of the St. Lawrence Great Lakes," Paper Presented at the symposium on "Recent Trends in Ecological Research in the Great Lakes," Ecological Society of America, American Association for Advancement of Science meeting, Cleveland, December 27: 240–254, online, https://aslopubs.onlinelibrary. wiley.com/doi/pdf/10.4319/lo.1965.10.2.0240.

Bogue, Margaret Beattie (2000) *Fishing the Great Lakes: An Environmental History, 1783–1933* (Madison, WI: University of Wisconsin Press).

Dove, Alice and Steven C. Chapra (2015) "Long-term Trends of Nutrients and Trophic Variables for the Great Lakes," *Limnology and Oceanography* 60, 696–721.

Kehoe, Terence (1997) *Cleaning Up the Great Lakes: From Cooperation to Confrontation* (DeKalb, IL: Northern Illinois University Press).

Kehoe, Terence (1992) "Merchants of Pollution? The Soap and Detergent Industry and the Fight to Restore Great Lakes Water Quality, 1965–1972," *Environmental History Review* 16 (Autumn): 21–46.

McGucken, William (1991) *Biodegradable: Detergents and the Environment* (College Station, TX: Texas A&M University Press).

McGucken, William (2000) *Lake Erie Rehabilitated: Controlling Cultural Eutrophication, 1960s–1990s* (Akron, OH: University of Akron Press).

Okun, Daniel A. (1972) "Phosphates in Detergents—Bane or Boon?" *Boston College Environmental Affairs Law Review* 2: 64–79.

Read, Jennifer (1996) "Let Us Head the Voice of Youth": Laundry Detergents, Phosphates and the Emergence of the Environmental Movement in Ontario," *Journal of the Canadian Historical Association* 7: 227–250.

Scherber, Annette Mary (2018) "'Clean Clothes vs. Clean Water": Consumer Activism, Gender, and the Fight to Clean Up the Great Lakes, 1965–1974,' (Unpublished MA Thesis, History, Indiana University, August).

Shear, Harvey, (2006) "The Great Lakes, n Ecosystem Rehabilitated, ut Still Under Threat," *Environmental Monitoring and Assessment* 133: 199–225.

18 The Fluoride Controversy

The recognition that water could be a medium for disease goes back at least to 1854 when Dr. John Snow surmised that a polluted well near Broad Street in London might be the source of cholera. Allegedly breaking the pump handle ended the scourge. Exactly what in the water caused disease awaited the discovery of bacteriology through the work of Louis Pasteur, Robert Koch, and others in the 1880s. By the turn of the century bacteria—or germs—were recognized as disease vectors in water, and thus scientists, sanitarians, and engineers began a search for ways to treat water supplies.

Seeking to mitigate against several forms of the disease in the early twentieth century led to introducing chemicals into water, especially chlorine. For many years before then (and in some cases afterwards), engineers were convinced that dumping industrial waste such as phenol into flowing water would disinfect it if the streamflow was strong enough. Such an idea seemed to be out of line with the bacteria—or germ—theory of disease. In almost every instance, however, the idea that additives to water offered a means to uphold its purity—or at least its potability—met with debate. Adding chemicals to water especially raised questions about safety. Was the cure worse than the disease?

The fluoridation controversy beginning in the 1940s revitalized the age-old concern over water purity. In this case, adding a chemical—fluoride—that was not meant to fight water-borne disease but to aid in the reduction of tooth decay. This was preventive medicine instead of disease eradication. The controversy itself revolved around trust in science and potential government overreach. To a much lesser degree, whispers of water adulteration in a Cold War context elicited fears about communist infiltration of our water supplies.

In the first decade of the twentieth century, water treatment made important strides. Chlorination was well established and experimentation with copper sulfate to control algae was underway. Leading sanitarians had been able to convince city leaders that typhoid fever and related diseases were preventable through a combination of filtration and treatment. Water-supply specialist Allen Hazen purportedly stated that since typhoid was preventable, for every death from the disease someone should be hanged. George A. Johnson, a consulting sanitary engineer, also fervently argued that "a person who knowingly takes into his mouth the excrement of another human being

DOI: 10.4324/9781003041627-25

is certainly defective. A community having a public water supply known to be contaminated by the sewage of other cities, and using it without even attempting its purification, is certainly a victim of defective civilization" (Melosi, 2008, 94).

Few major proponents of water treatment claimed that filtration was the only answer to disease prevention, but contemporary statistics indicated that it was especially effective in controlling typhoid fever. The results often were so dramatic that people lost sight of the idea that filtration's primary purpose was reducing turbidity and removing suspended matter.

Disinfection became a complement to filtration and promised to play an even greater role in removing bacteria from the water supply. Ancient civilizations disinfected what they believed to be tainted water through boiling or by using copper vessels. A professor at the Stevens Institute in New Jersey patented a process for the chlorination of water in 1888. But even before chlorination became popular, sewage had been chlorinated in England, France, and parts of the United States.

Bleaching powder was first used to chlorinate water in Austria in 1896. The next year, chlorine gas was applied to a test filter at Adrian, Michigan. The first continuous-use water chlorination plant was constructed in Belgium in 1902. The initial use in the United States took place at Bubbly Creek Filter Plant at the Union Stock Yards in Chicago in 1908. Jersey City became the first community to have its water supply chlorinated soon after the Bubbly Creek experiment. In 1909, liquid chlorine was produced, which provided a much easier method of dispersal.

A dramatic decline in typhoid fever rates followed the use of chlorine in many locations. Statistics from cities utilizing hypochlorite in their water offered some striking evidence of the change. Yet, despite the optimistic reports, there remained concern among some citizens about "doping the water" with chemicals. While this fear had been greater before the turn of the century, it never completely died out and thus chlorination was limited in use by 1920. An *American City* survey for that year showed less than half of approximately 1,000 cities using chlorine or some other disinfectant in their water supplies.

From a public perspective, water supplies had to be plentiful, cheap, and safe. Availability of abundant water at a reasonable cost varied from region to region, but assessment of water quality was more standardized by the late 1920s. The interwar years witnessed a better understanding about what constituted water pollution, and the process of purifying and treating water was significantly refined. There was more complete preliminary clarification of water prior to filtration, better mechanical filtration, better controlled use of chlorination, new procedures for reducing odors and tastes, attention to corrosive elements in water, broader use of aeration, and progress in softening water. The debate continued, nevertheless, on when to filter water versus when to chlorinate, and when sewage treatment was preferable to filtration (Melosi, 2008, 94–95).

Chlorination continued to receive major attention, especially because of periodic outbreaks of epidemics such as typhoid fever. Yet, only about one-third of all waterworks employed chlorination by 1939. Some resistance to chlorination came as a result of taste and odor problems. Industrial wastes, especially phenol, reacted with chlorine to produce a bad taste in the water. Chloramine, used in Great Britain, began to gain favor in the United States in the 1930s as a replacement for chlorine. It proved to be a better bactericide and it curbed bad odors and tastes. In 1924, activated carbon was used as a filtering agent and went on to gain great popularity. By 1943, almost 1,200 plants in the United States were using activated carbon for odor control.

In 1966, some form of chlorine was employed by 99 percent of all municipalities that chemically disinfected their supplies. Concern arose over the ability of chlorination to keep up with increasing water demand; and tests showed that chlorine was not effective against all microorganisms, at least in the concentrations used in existing waterworks systems.

Sometimes other chemicals were added to the water supply to protect against various threats to health or for other purposes. Lime treatment became fashionable for softening water. Laundries preferred the soft water because it took less soap to clean clothes. Boiler plants used soft water to prevent encrustation in boiler tubes. However, soft water was corrosive to iron and steel, which could adversely affect miles of pipe, and thus demand varied from location to location.

Another water additive, fluoride (one of the most abundant elements found in Nature), offered health benefits via the water supply. It was not used as a protective or preventive treatment such as filtration, disinfection, or coagulation, but for the purpose of reducing dental caries (tooth decay)—a chronic disease especially prevalent in children. Some water sources contained natural fluoride, but for those that did not, it could be added. The role of fluoride in preventing dental caries was under discussion in Europe in the nineteenth century. Proposals to fluoridate municipal water supplies, however, originated in the United States.

In about 1902, dentist Frederick S. McKay began to study mottling—brownish staining—on the teeth of some residents in Colorado Springs. While this "Colorado brown stain" caused no physical harm it was unsightly and appeared to originate in the water supply. Although some residents attributed it to eating too much pork or drinking bad milk. In 1909, McKay surveyed approximately 3,000 children from the Pikes Peak region of Colorado and found that 87.5 percent of them were suffering some degree of enamel staining (fluorosis). Coincidently, the kids had few cavities. He eventually discovered that the mottling also occurred in places in Italy, Portugal, and other parts of the Rocky Mountains, Texas, Arkansas, and elsewhere in the United States.

After years of investigation by McKay and others, two separate studies in 1931 concluded that excessive amounts of naturally occurring fluoride in local water supplies caused the distinctive mottling. A dental surgeon with the U.S. Public Health Service (USPHS) confirmed that mottled teeth were particularly resistant to decay. Over the next six years, studies of children not only in Colorado, but also in Ohio, Illinois, and Indiana, led to the conclusion that fluoride in small concentrations (about 1 part per million) significantly reduced cavities without causing stains in the mouths of most children. Henry Churchill, the chief chemist for the Aluminum Corporation of America (ALCOA), produced the first analytical evidence connecting brown stain and fluoride in water supplies, through his work on the ALCOA company town of Bauxite, Arkansas.

In 1944, the City Commission of Grand Rapids, Michigan, voted to add fluoride to its public water supply. In 1945, it became the first city in the world to introduce a fluoride concentration of 1 ppm (powdered sodium fluoride) to its drinking water. Three other case studies were soon added to the list: Brantford, Ontario (1945); Newburgh, New York (1945); and Evanston, Illinois (1947).

Madison, Wisconsin, dentist John Frisch and the state dental-health officer, Francis Bull, called for immediate action in their state. Between 1946 and 1950, Frisch traveled across Wisconsin promoting fluoridation. He claimed, all the "rabid skeptics" came back "fluoride enthusiasts of the first magnitude" (McNeil, 1985, 144). Despite the slightly overstated case, by January 1950, 50 Wisconsin communities had fluoridated water.

In that year, proponents of fluoridation, including the U.S. Public Health Service and the American Dental Association, were totally on board. Endorsements for fluoridation also came from the American Medical Association and the American Public Health Association (eventually Canadian medical, dental, and public health bodies joined in). Supporters considered it a safe, efficient, and cost-effective way to reduce dental decay—a major public health achievement. Medical historian Christopher Sellers stated, "On January 25, 1945, the first fluoride-loaded drops of drinking water splashed through the faucets of Grand Rapids, Michigan. They bore the rising hopes of a handful of public health officials that fluoridation would prove a 'magic bullet' remedy against tooth decay" (Sellers, 2004, 182).

Figure 18.1 An anti-fluoride sticker on an underpass in Wanniassa, Australia, April 2013. Creative Commons Attribution-Share Alike 3.0. Wikipedia.

By 1951, public acceptance of adding fluoride to water supplies was on the rise in the United States. In that year, two councils of the American Medical Association and a special committee of the National Research Council issued statements declaring that there was no evidence of toxicity in the process. Canadian medical and public health associations lagged in their endorsement because they did not believe there was enough known about the long-term effects of fluoride. However, in 1953, the Canadian Dental Association endorsed fluoridation in Canadian water supplies followed soon after by the Canadian Medical Association. During the 1950s and 1960s, evidence of the value of water fluoridation continued to grow. By the early 1960s, school children showed significant improvement in dental health in Philadelphia, one of the earliest large cities to adopt the practice. Many state governments passed laws allowing local authorities to add fluoride to their water. For its part, the Public Health Service began holding annual conferences on fluoridation.

Proponents—because of optimism in their findings or aspirational projections—claimed that there was no legitimate scientific opposition to fluoride, criticism only coming from the fringes of society, from unorganized cranks and quacks. Such declarations understated and underestimated the anti-fluoridation response over the years, especially since some scientists legitimately were concerned about supporting an early rush to fluoridation. While fluoridation of water supplies remained on solid research ground, there were cracks (sometimes small cracks) in the debate among scientists and researchers. By the mid-1940s, as historian Catherine Carstairs stated, "An increasing number of studies had demonstrated that fluoride was effectively excreted by the body and the apparent healthfulness of naturally fluoridated communities relieved many people's concerns. But scientific consensus had not yet been achieved" (Carstairs, 2015, 1561).

Evidence on skeletal fluorosis (a disease caused by excessive accumulation of fluoride leading to weakened bones) from Southern India published in 1937 did not jive with American studies and was dismissed in the United States as not applicable. Additional research found mild mottling caused by low levels of fluoride in the water. Biochemists at the University of Wisconsin opposed fluoridation of Madison's water supply in 1947, arguing that fluoride tablets would be a better solution. There also was a claim that some scientists at the USPHS reluctantly endorsed fluoridation in 1950 because of pressure from state dental directors.

Anti-fluoridation responses (in Canada as well as the United States) quickly left the realm of medicine with the first applications of fluoride in Grand Rapids. In some respects, opposition to fluoridation was similar to that raised against chlorination, that is, skepticism about any compound added to a natural water supply. In the Cold War atmosphere of the 1950s, fear of radioactive contamination added to concern about the purity of national water supplies and the possible introduction of foreign substances into the water. The fight over fluoridation soon became the most controversial water-treatment issue in the postwar years.

In a variety of towns and cities lawsuits were posted and even national bills sought to ban fluoridation. More than 20 Texas towns and cities, on the verge of fluoridating its water, postponed the decision. Some cities dropped the practice after an initial trial, such as Stevens Point, Wisconsin. Alexander Y. Wallace, who led the backlash there, called fluoride "a poison" and referred to Frisch and his supporters as "foreigners" and faddists (McNeil, 1985, 145–146). Between 1953 and 1963, fluoridation was halted at 60 water-supply systems, although 26 eventually reinstated it. Opposition to fluoride led to hearings in Washington as early as 1951.

According to historian of the fluoridation controversy Donald R. McNeil, "Before Stevens Point, there was no organized opposition to fluoridation. There were only scattered cries of alarm from people (like Wallace) who had earlier opposed the pasteurization of milk and other government-imposed measures" (McNeil, 1985, 146). National networks of antis started to form. Some dentists and scientists, chiropractors, natural health advocates, and others gained allies in a variety of organizations from the arch-conservative John Birch Society and the Ku Klux Klan to Christian Scientists and *Prevention* magazine. Various groups began sharing information and tactics to oppose fluoridation.

Into the 1950s, some criticized city-sponsored fluoridation as an assault on personal freedom, or unnatural "mass medication" forced on citizens without their consent. Some simply referred to sodium fluoride as "rat poison," despite its dose level. Seattle radiologist Dr. Frederick B. Exner said the practice "violates the most sacred rights of God and man" (Melosi, 2008, 184). But there were legitimate questions as well: What were "safe" levels of fluoride? Wasn't fluoridation wasteful, since only a small amount of the water supply is ingested?

Probably the most outlandish claim in the McCarthyite Era was that fluoridation was a communist scheme to kill or weaken Americans. Golda Franzen, a housewife in San Francisco, became a leading promoter on fluoridation as a "Red conspiracy." In Stanley Kubrick's 1964 black comedy, *Dr. Strangelove,* the deranged General Jack Ripper—who had independently unleashed a bomber wing with nuclear warheads on the Soviet Union—tells his executive officer that fluoridation was "the most monstrously conceived and dangerous communist plot we have ever had to face" and sapped "our precious bodily fluids." As historians, Carstairs and Rachel Elder stated, "While Kubrick's film mocked Ripper, arguments that fluoride would be used to make the population docile or that it provided a potent tool to communists wishing to poison the water supply were common in the 1950s in both the United States and Canada" (Carstairs and Elder, 2008, 363–354).

Figure 18.2 General Jack D. Ripper (Sterling Hayden) in Stanley Kubrick's 1964 film, Dr. Strangelove. In his mind,fluoridation was "the most monstrously conceived and dangerous communist plot we have ever had to face"and sapped "our precious bodily fluids." Creative Commons Attribution 2.0. Wikimedia.

In the long run, the support of key public health and scientific groups, and the positive results on dental health, led to increasing use of fluoridation. During the 15-year fluoridation study which began in Grand Rapids in 1945 researchers monitored the rate of tooth decay among approximately 30,000 school children. After 11 years, it was announced that the rate of caries among the children born after fluoride was added to the water supply dropped by more than 60 percent. In 1951, more than 360 communities had adopted the process; by 1968, more than 4,000. After ten years of use in Newburgh, New York, there was 41 percent less tooth decay among 16-year-olds than in Kingston, where no fluoridation took place. Among a 6-to 9-year-old group, the results were more dramatic with 58 percent less tooth decay.

Water purity and related health issues continue to be key concerns to this day, including debates over the use of chlorine and the introduction of fluorine into water supplies. Although apprehension over a range of environmental risks threatening the nation's water supplies and complicating the process of waste-water treatment persists, outbreaks of water-borne diseases continue to lose ground in the United States and Canada. Diseases such as cholera and typhoid fever have been virtually eliminated, and the few outstanding disease outbreaks primarily occurred in small community or noncommunity systems. Periodically, a large outbreak did arise, as in the case of cryptosporidiosis (an infection caused by parasites) in Georgia in 1987 affecting 13,000, or a major attack of gastroenteritis (irritation or inflammation of the stomach and small intestines) affecting 403,000 in Milwaukee in 1993. Contamination in distribution systems and inadequately disinfected water were significant causes of outbreaks. However, as water-borne disease outbreaks became increasingly rare, water pollution by a raft of chemicals or deterioration in water pipes and in filtration and treatment facilities proved to be serious problems. (See, for example, Essay 23 in this book on Flint's water crisis.)

Disinfection remains one of the most important steps in water treatment. In 1970, chlorine accounted for 95 percent of the disinfectants used to treat potable water in cities. Aside from increasing costs, chloramines were believed to have a limited impact on viral agents in the water, and thus attention turned to alternative disinfectants such as other halogens, bromine, and iodine. Chlorine disinfection also produced some undesirable by-products, including trihalomethane (THM; solvents or refrigerants that may be carcinogenic).

To control THM, officials took a new look at ozone treatment because of its high germicidal effectiveness, its ability to combat odor, taste, and color problems, and its benign decomposition. There had been resistance to ozone because it was produced electrically and could not be stored, was difficult to adjust to variations in water quality, and was not universal in its action as a disinfectant. No consensus emerged to abandon chlorination, and the U.S. Environmental Protection Agency (EPA) still believed that it was the most effective commonly used additive for controlling water-borne diseases.

Despite its impressive record, chlorination came under severe criticism. A 1974 report alleged a causal relationship between the use of Mississippi River water (containing chlorinated sewage effluent) and the incidence of cancer. While the study's validity was later questioned, it led to an increased awareness of chemical contaminants in water supplies and the impact of chlorination. In April 1975, EPA reports of cancer-causing chemicals in the water supplies of 79 cities splashed on the front pages of many newspapers. Water scares resulted in cities such as New Orleans and Duluth, Minnesota.

In March 1976, the National Cancer Institute issued a report indicating that chloroform was carcinogenic. The EPA set the first THM limits in 1979. Health officials also monitored fluoridation use and practices. In 2016, the Department of Health and Human Services released new recommendations for fluoride levels in drinking water, updating those in place since 1962 because more sources of fluoride other than drinking water were available.

Fluoridation suffered its own scrutiny as debates continued over questions of effectiveness, potential risks, and government decision-making. In 1990, fluoride was being added to the water supplies of 57 percent of the American people and was endorsed as a tooth-decay preventative by most major health organizations. In the late 1980s, over 300 million people worldwide drank water with natural or artificially maintained levels of fluoride; in 2020, the number was over 400 million.

Other groups such as the National Health Action Committee, the Fluoridation Action Network, and the Safe Water Foundation actively challenged community water fluoridation and fluoride mouth-rinse programs in the schools. Those groups were not opposed to fluoride per se, but more concerned with involuntary mass medication, claimed that fluoride had become available in toothpaste and rinse, thus adding it to water supplies was unnecessary. The decision to use these products or not, they argued, should be left to the individual not through passive intervention.

Efforts to change policies at the community level probably led to the most intense reactions. In 2013, Portland, Oregon's City Council, voted to fluoridate its water after many years of not doing so. Anti-fluoride activists gathered 20,000 signatures to put the issue on the ballot. The plan to fluoridate the water failed by a large margin.

As McNeil stated, "The fluoridation question is almost tailor-made for endless controversy in a free-wheeling democratic society. Fluoridation, unlike chlorination, is not a life-or-death matter. Its scientific rationale has little emotional appeal. When it comes to a vote, fluoridation is largely a symbolic issue" (McNeil, 1985, 153).

Proponents charged the opposition with employing scare tactics, spreading half-truths, and even attempting to link fluoridation with aging and AIDS. But McNeil, writing when he did in 1985, undervalued the medical component of the debate. Some research studies more recently have raised doubts about the ability of fluoride to curb tooth decay, suggesting that there were no appreciable differences in groups exposed to fluoride and those who were not. Advocates of fluoridation celebrated what they view as substantial decline in cavities in teeth from 1950 to 2000, while some controlled studies indicated that fluoridation reduced cavities by only 15 to 35 percent—far less than the claim of 60 percent or more by proponents. However, proponents argue that the lower percentiles reflect a more recent population with already improved dental health. The debate goes on.

Some concern also mounted worldwide about incidences of dental fluorosis—in the most extreme cases leading to brown stains and pitting of the teeth. More serious were charges—not proven—about the negative impact of fluoridation on pregnant women and the potential correlation between cancer and fluoridation.

To be fair, proponents of fluoridation in the early years rushed to judgment on sometimes incomplete or poor-quality research, producing legitimate skepticism among some in the scientific and health communities. As Carstairs concluded, "Although water fluoridation undoubtedly did improve the dental health of many children in the 1960s and 1970s, fluoride proponents were perhaps too hasty in declaring that community fluoridation was the best (or only) solution for dental decay"(Carstairs, 2015, 1567).

Sellers argued, "The science justifying fluoridation, like the practice itself, has remained predominantly American" (Sellers, 2004, 183). He asserted that American promotion of fluoridation arose in a rather closed system. Scientists outside the United States and Canada (and some within) had concerns about fluoridation beyond its role in fighting tooth decay—more skeptical of its "universal value" or about its environmental ramifications writ large. This was compounded by the fact that natural concentrations of fluoride varied by climate and region globally. For example, India had to grapple with excess amounts of fluoride in its drinking water, which posed a different set of issues (including widespread problems of skeletal fluorosis) than in the United States where fluoridation proponents were trying to establish the appropriate amount in water supplies lacking fluoride.

At the heart of the fluoride debate in the United States and Canada was a concern over tinkering with Nature—adding a substance to drinking water, which was not there already or at least in a predetermined ratio. Advocates touted the value of improving the health potential of the water source with the additive, while detractors raised a flurry of objections (from a variety of perspectives) asking why it was necessary to meddle with the natural order of things.

Many controversies surrounding science often come down to these fundamentals. Resolution is difficult, or impossible, when the sides start from different places or lack trust in an opposing side. And too often such debates devolve into confrontation between two intractable groups with absolute rights and wrongs at stake. Water, in the case of the fluoride controversy, was the battleground rather than the arbitrator.

Notes and Further Reading

Aoun, Antoine, et al. (2018) "The Fluoride Debate: The Pros and Cons of Fluoridation," *Preventive Nutrition and Food Science* 23 (September): 171–180.

Carstairs, Catherine (2015) "Debating Water Fluoridation Before Dr. Strangelove," *American Journal of Public Health* 105 (August): 1559–1569.

Carstairs, Catherine and Rachel Elder (2008) "Expertise, Health, and Popular Opinion: Debating Water Fluoridation, 1945-1980," *Canadian Historical Review* 89 (September): 345–371.

Garfield, Eugene (1986) "Fluoridation, 'Texas Teeth,' and the Great Conspiracy. Part 1, The Issues," *Essays of an informed Scientist* 12: 3–9.

Hicks, Jesse (2011) "Pipe Dreams: America's Fluoride Controversy," *Distillations*, June 24, Science History Institute, online, https://www.sciencehistory.org/distillations/pipe-dreams-americas-fluoride-controversy.

McNeil, Donald R. (1985) "America's Longest War: The Fight over Fluoridation, 1950—," *Wilson Quarterly* 9 (Summer): 140–153.

McNeil, Donald R. (1957) *The Fight for Fluoridation* (New York: Oxford University Press).

Melosi, Martin V. (2008) *The Sanitary City: Environmental Services in Urban America from Colonial Times to the Present* (Pittsburg, PA: University of Pittsburgh Press; abridged ed.).

Sellers, Christopher (2004) "The Artificial Nature of Fluoridated Water: Between Nations, Knowledge, and Material Flows," *Osiris*, 2nd Series, 19: 182–200.

Till, Christine and Rivka Green (2020) "Controversy: The Evolving Science of Fluoride: When New Evidence Doesn't Conform with Existing Beliefs," *Pediatric Research* May 22, online, https://pubmed.ncbi.nlm.nih.gov/32443137/.

19 Hurricane Hazel: In Canada

The most dramatic and potentially most destructive events associated with water are hurricanes or cyclones. They are generally allied with tropical weather and their impact primarily focuses on coastal communities. Since humans cannot control the formation of hurricanes/cyclones or determine their paths, the best we can do is improve our ability to understand and track them effectively and then prepare for them. Neither of these things occurred in 1954, when Canada faced its worst and most famous hurricane and Southern Ontario incurred its worst natural disaster.

This is not a typical story hurricane story. In October, Hurricane Hazel formed in the tropics, stalked through the interior of the United States, and crossed over the Great Lakes. There it merged with a cold front and became an extratropical storm that raised havoc on an unsuspecting Toronto. Hazel I became Hazel II. Such events are rare but do suggest an important degree of unpredictability in dealing with weather and the potency of water—and all that came with it—as a destructive force.

Hurricanes, which are severe tropical storms developing over warm ocean waters, can produce powerful winds, severe flooding, and strong storm surge. They form over the ocean, often beginning as tropical waves, that is, low-pressure areas traveling through the tropics rich in moisture. As warm ocean air rises into the storm and rotates, low pressure then forms underneath causing more air to rush in. The air rises and cools which forms clouds and thunderstorms. In the clouds, water condenses and becomes droplets which release more heat that powers the storm. Hurricanes take energy from the warm ocean water to become stronger but weaken over land. A tropical storm is said to be a hurricane when it reaches a wind speed of 74 miles per hour (mph). According to the National Ocean Service, "During just one hurricane, raging winds can churn out about half as much energy as the electrical generating capacity of the entire world, while cloud and rain formation from the same storm might release a staggering 400 times that amount" (National Ocean Service, 2021).

Hurricane systems normally originate off Africa's coast and then travel across the Atlantic Ocean. Intense storms called typhoons or tropical cyclones form over the North Atlantic, Northern Pacific Ocean, Southwestern Pacific, and

DOI: 10.4324/9781003041627-26

Indian Ocean. Because of the earth's rotation, Northern Hemisphere hurricane winds circulate counterclockwise, while cyclones in the Southern Hemisphere rotate clockwise. Atlantic hurricanes threaten the eastern part of the United States from approximately June through October. The Pacific typhoon season officially begins May 15 and extends into late November.

Many of these storms spawned in the ocean never reach land, but those that do can be dangerous and deadly. We most often measure the devastation of these so-called "natural disasters" by their impact on lives and property but tend to forget that human settlements—by choice or by circumstance—have placed themselves in harm's way. As historian Ted Steinberg argued, "Natural disasters have come to be seen as random, morally inert phenomena—chance events that lie beyond the control of human beings. In short, the emphasis has been on making nature the villain." So-called "acts of God," he added, have become "little more than a convenient evasion" where the human element in the death and destruction of these storms tends to be unrecognized or underplayed (Steinberg, 2000, xxv).

In recent years, debate over the impact of climate change (or more narrowly referred to as global warming) on the frequency and size of hurricanes/cyclones has been fierce. Scientists in general are not certain as to whether climate change will lead to increased numbers of such storms. However, many climate scientists believe that warmer ocean temperatures and sea level rise intensify them substantially. For example, studies show that there has been an increase in extreme hurricane activity in the North Atlantic since the 1970s. According to historian Anthony N. Pena and environmental studies professor Jennifer S. Rivers, "Worldwide, the destruction caused by tropical cyclones has intensified during the last thirty years [about 1983 to 2013], with the number of category 4 and 5 cyclones [above 135 mph] nearly doubling since 1970" (Penna and Rivers, 2013, 226).

According to the National Weather Service, "Hurricane Hazel was the deadliest hurricane of the 1954 hurricane season and is the strongest and only Category 4 hurricane to ever hit the North Carolina coast" (National Weather Service, 2021). The first observation that a tropical storm had formed occurred on October 5 in an area 50 miles east of the island of Granada in the Windward Islands. What became Hurricane Hazel moved westward over the Caribbean Sea through October 8, then turned northward. By the next day, Hazel had become a Category 4 storm with maximum winds at 135 mph. On October 11, it crossed Western Haiti (the death toll estimate of 400–1,000 people), then passed over the Southeastern Bahamas (six deaths) on October 13.

Moving northwestward on the 14th, its winds increased to 150 mph. Originally thought to remain offshore, Hazel made landfall on the border between North and South Carolina on the 15th. Several communities were hit with high wind and storm surge along the North Carolina shoreline at more than 18 feet. Coinciding with a lunar high tide, the storm nearly destroyed Garden City, South Carolina, and did major damage elsewhere. The storm caused massive evacuations, 19 deaths in North Carolina, wrecked every

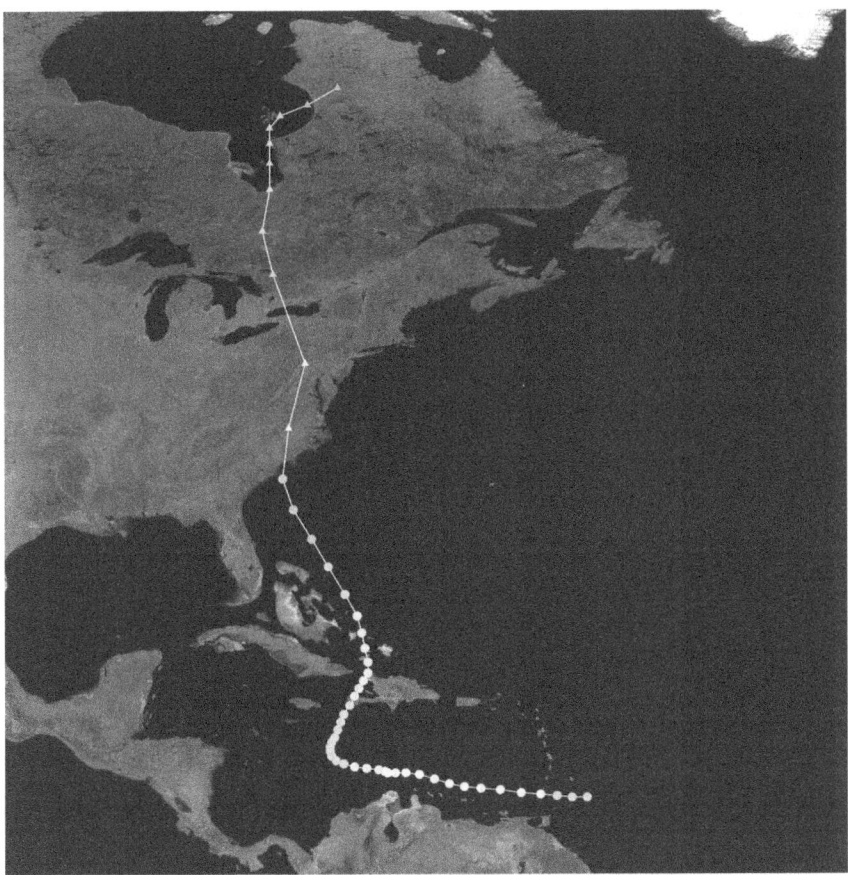

Figure 19.1 Track map of Hurricane Hazel of the 1954 Atlantic hurricane season. The points show the location of the storm at 6-hour intervals. Storm type: Public domain. Wikimedia.

pier along 170 miles of coastline, destroyed 15,000 homes, and damaged 39,000 more. The storm tracked northward near Wilmington, North Carolina, passing over the western suburbs of Washington, D.C., and continuing across the Appalachian Mountains (where it dropped 10 inches of rain) into Pennsylvania and then to Canada. Fierce winds were felt in Virginia, Maryland, Pennsylvania, Delaware, New Jersey, and New York. Overall, fatalities in the United States reached 95.

On October 15, the remnants of Hazel (still powerful) moved northward toward the Great Lakes. Typically, tropical hurricanes tend to dissipate as they move into northern land areas. This one did not, confounding weather trackers. As the decaying Hazel continued to move northward over Lake Ontario, an incoming cold front from the west collided with the weakened storm rapidly strengthening it into an extratropical cyclone packing wind gusts of 90 mph and generating plenty of rain. Hazel II then headed into Southeastern Ontario and Southwestern Quebec. An article in *The Canadian Encyclopedia* noted: "As Hazel approached Ontario, the Dominion Weather Office tracked its erratic path as well as the predictions by the American weather service [U.S. meteorologists assumed Hazel would dissipate before entering Canada], and issued warnings to broadcasters. However, few people there had experience with hurricanes and were unaware of how to prepare, leaving them vulnerable to the storm's power" (Marsh, 2015).

The idea—put forth by American forecasters—that Hazel would blow itself out before reaching Canada did not take account of the colliding cold front and the possibility of heavy rains. Fred Trumbull, head of the weather office in Malton (a neighborhood in the northeastern part of the city of Mississauga, Ontario) issued a statement suggesting that Toronto could experience the heaviest rain it ever recorded. Overall, what was broadcast to the citizenry was low-key and nonalarmist, and likely appeared little different than what they had heard over the years about bad rainstorms. No civil-emergency entity was alerted; in fact, Toronto was without any civil-defense unit. Locals warned about the considerable rain coming had no way of knowing the scale of what might occur and went about their normal routines.

According to a brief chronicle of Hazel in Canada: "There is much dispute about whether the residents of Toronto and the surrounding areas were forewarned about the hurricane. Many people do not remember having heard an announcement on the radio, and certainly not on their televisions" (Gifford, 2004, 23). But what news would have registered in a place where hurricanes were rare? As author Betty Kennedy concluded, "...to most Canadians a hurricane belongs to other people, other places" (Kennedy, 1979, 1).

Toronto was the area hardest hit by Hazel II. It proceeded to pound the metropolitan region with 68 mph winds and more than 11 inches of rain in 48 hours. Making matters worse, in the weeks before the storm the Toronto area had received higher-than-normal rainfall that already dangerously saturated the water table. Estimates suggested that more than 90 percent of Hazel's rain ran off into rivers and creeks, raising water levels to about 26 feet. Areas northwest of Toronto—Snelgrove and Brampton—received the most rain. The metropolitan area's infrastructure was not built to withstand the heavy flooding that occurred, and more than 50 bridges were destroyed or badly damaged and numerous roads and railway lines north of the city were washed out. Passenger trains were knocked off their tracks.

Woodbridge, northwest of Toronto, was the first community to be inundated by the rising waters. Lieutenant John P. Connor and members of the Royal Canadian Sea Cadet Corps manned two 27-foot whalers to check houses for stranded people. Connor recalled:

> The most memorable event at this point was rescuing an 82-year-old bed-ridden grandmother—mattress and all, and her son who, two days

before, had been discharged from hospital, having been confined for pneumonia…He stayed on his front porch all night, and with the aid of a broomstick, he personally rescued twenty-seven cats and fourteen dogs. All were loaded aboard the whaler and the menagerie huddled together for warmth…When we reached high ground near the Fire Hall, our cargo of cats and dogs were last seen heading for still higher ground—but still together.

(Kennedy, 1979, 79)

Holland Marsh, a fertile 10,000-acre flatland located in a bowl-like valley south of Lake Simcoe, flooded slowly giving people time to escape to the town of Bradford, which was located on a hill. Nearby Highway 400 was swamped with more than 19 feet of water in some places. Draining the marsh was a very long process taking until mid-November. Crops there (it was an important vegetable growing area) had been harvested but had not been transported and were submerged or washed away.

The largely deforested Humber River in the west end of Toronto, which was prone to flooding, caused the most destruction because of flash floods. The *Toronto Star* reported the police were informed that the current was so strong that launching a rescue boat was folly since "nothing can make it and anyone in it will be killed for sure." A team of five volunteer firefighters died trying to help stranded motorist when their fire truck was washed away.

In all, 81 Canadians died as a result of Hazel II, 35 of them along Raymond Drive, which ran parallel to the Humber. Fourteen homes, many with the occupants inside, were swept away. Because of the unprecedented rise of the river, residents were caught off guard and had not evacuated, which led to the high death toll. One volunteer firefighter was quoted in the *Toronto Star* (October 14, 1984): "I felt so helpless, but there was nothing I could do, nothing anybody could do. The water was so deep, up to our chins, and all the firemen were weighed down by clothing and boats and equipment." The damage was so severe along Raymond Drive that the neighborhood later was converted into a park. The Etobicoke Creek, further west, overflowed its banks at Long Branch near Lake Ontario, which resulted in seven deaths. The village also was converted into a park ("Hurricane Hazel, 1954," 2016).

Toronto experienced the worst flooding in more than 200 years. Aside from the deaths, infrastructure damage, and loss of houses and trailers, about 1,800 families were left homeless. The total cost of destruction was estimated at $100 million (about $1 billion in today's dollars). The city's landscape was permanently changed and the need for new plans to manage the regional watersheds was evident.

On the day after the storm, the *Globe and Mail* reported that at Fair Glen Crescent where 20 homes had been destroyed "Grotesque, lopsided houses sagged or lay in every imaginable position…Tops and hoods of cars protruded

Figure 19.2 Damage from Hurricane Hazel along the Humber River in Woodbridge, November 1954. James Victor Salmon. Public domain. Wikimedia.

above the water, marking the place where Fair Glen Crescent bends and follows the river-front [on the Humber]" (Gifford, 2004, 49).

In the immediate aftermath, local service clubs, women's organizations, the Salvation Army, the Boy Scouts, the Red Cross, and others served meals to the homeless and helped in any way they could. People provided shelter, clothing, and other items for victims who needed to pull their lives back together. Eight hundred army troops were summoned to help with the cleanup in Toronto. Members of the Navy provided boats and men. Financial assistance came from the local, provincial, and federal governments at various levels of support. The Hurricane Relief Fund, private donations from big companies such as Ford Motor Company, and smaller ones, such as Laura Secord Candy Shops, and aid from churches helped the cause. Insurance companies, for their part, set up offices in New Toronto, Woodbridge, and Newmarket to handle claims. But in many cases, damage because of flooding was not covered in most policies.

With respect to longer-range planning, conservation authorities, local municipalities, and the provincial government sought to develop plans for flood control and water conservation in the wake of future extreme weather events. The Conservation Authorities Act allowed conservation authorities to acquire and regulate vulnerable areas for recreation and conservation purposes in places such as those along Raymond Drive. (The act was controversial in that some homeowners—especially mobile homeowners—complained about the financial compensation for expropriated land.)

In 1957, the Metropolitan Toronto and Region Conservation Authority was formed out of several regional authorities and commenced planning for the development of large dams, reservoirs, flood-control channels, and for an erosion-control program. Since Hazel II regulations restricted new development in flood plains. A 2006 study of disaster relief after Hazel concluded, "A tropical storm traveling through the province of Ontario was a relatively rare event, yet ultimately government officials did not respond to the Hazel disaster as a random, chance event. Instead, the conservation movement and local authorities pressured governments to see the hurricane flooding not as a natural disaster, but as a tragedy, which human decisions had helped precipitate, and which, in the future, human decisions might alleviate" (Robinson and Cruikshank, 2006, 37).

The chances of Ontario Province experiencing major impacts from a hurricane was calculated at one percent in any given year (Gifford, 2004, 19). Yet Hazel defied the odds. Given its destruction across the United States and Canada, the name was retired from use. There have been much worse hurricanes over the years in the United States, with much more cataclysmic results—at least in terms of scale. What made the impact of Hazel II in Canada so tragic in 1954 was how it blind-sided officials and citizens in the Toronto area because it was so unexpected.

If climate scientists are correct, storms we face in the future may confound us even more in many places and become even less predictable as Hazel was in Canada. For example, in August/September 2017 Hurricane Harvey produced an 8-day rainfall in Houston and environs resulting in 60 inches of rain in some locations about 7 or 8 more than the yearly average—and massive, widespread flooding. Because of the deluge, the National Weather Service had to add two more levels to its rainfall maps just to reflect what might become the new normal.

Preparation is everything, but for what? remains unclear. High winds? Flooding? Storm surge? All three? Contingency planning is necessary, but experts and government officials are faced with taking into account not only past trends and possible futures, but also the uncomfortable reality that humans have plunked down their settlements on such a scale, and sometimes in such random fashion, that adequate preparation can seem almost impossible, if not folly. Hurricane Hazel may have been an unpredictable event for Toronto and environs, but maybe we need new and more comprehensive contingency measures to help us survive such events.

Notes and Further Reading

Gifford, Jim (2004) *Hurricane Hazel: Canada's Storm of the Century* (Toronto: The Dundurn Group).

"Hurricane Hazel, 1954," (2016), October 18, online, https://ininet.org/hurricane-hazel-1954.html.

Kennedy, Betty (1979) *Hurricane Hazel* (Toronto: Macmillan of Canada).

Marsh, James H. (2015) "Hurricane Hazel," *The Canadian Encyclopedia* (August 20), online, https://www.thecanadianencyclopedia.ca/en/article/hurricane-hazel.

National Ocean Service (2021) "How Do Hurricanes Form?" National Ocean Service, National Oceanic and Atmospheric Administration, U.S. Department of Commerce, May 13, online, https://oceanservice.noaa.gov/facts/how-hurricanes-form.html.

National Weather Service (2021) "Hurricane Hazel, October 15, 1954," National Weather Service, National Oceanic and Atmospheric Administration, U.S. Department of Commerce, online, https://www.weather.gov/mhx/Oct151954EventReview.

Penna, Anthony N. and Jennifer S. Rivers (2013) *Natural Disasters in a Global Environment* (Chichester, West Sussex, UK: Wiley Blackwell).

Robinson, Danielle and Ken Cruikshank (2006) "Hurricane Hazel, Disaster Relief, Politics, and Society in Canada, 1954-55," *Journal of Canadian Studies* 40 (Winter): 37–70.

Steinberg, Ted (2000) *Acts of God: The Unnatural History of Natural Disaster in America* (New York: Oxford University Press).

Weese, Scott R. (2003) *A Reanalysis of Hurricane Hazel (1954)* (MS Thesis, Atmospheric and Oceanic Sciences, McGill University).

Part VII

The New Ecology

The emergence of the New Ecology and the modern environmental movement produced a paradigm in which, among other things, issues related to water were being viewed with different eyes—at least among scientists and environmental activists, if not governments and private businesses. The new environmental awareness, emerging after World War II and blossoming in the 1960s, had the potential to change our appreciation of the physical world, but still fell short on many occasions.

The basic concept of ecology revolves around "the relationship between the environment and living organisms," particularly the reciprocal relationship between the two (Knight, 1965, 2). The rise of ecology as a science coincided with the industrial era beginning in the early twentieth century. The New Ecology in practice was not an all-embracing environmental concept as much as a shift in emphasis toward different priorities. By the 1960s, ecology changed from a scientific concept to a popular one as the questioning of traditional notions of progress and economic growth were more frequent (Worster, 1977, 339–340). Rachel Carson's *Silent Spring* (1962), a grim warning of the dangers of pesticides, seemed to best capture the new spirit. Career ecologists made it clear that "respect for the biosphere, like respect for justice, must continuously have a place in law and government" (Scheffer, 1991, 4).

Ecology was a helpful blueprint for constructing national environmental policy, but it was not always embraced. In its simplest form, the New Ecology could guide people from the utilitarian conservationism of the past to an era emphasizing environmental quality and personal health and well-being.

As the essays in this section suggest—and some of those in earlier sections as well—such a commitment to ecological principles did not necessarily guide decision-making over crucial problems related to water.

The first water episode deals with the Ixtoc 1 oil spill in the Gulf of Mexico in 1979. It is historically significant because it was the largest spill in the world at the time. The spill took place in Mexican waters and had an immediate impact on the Mexican oil industry, and also on the beaches of Mexico and Texas. But the Ixtoc 1 spill was only one of many over the years including the massive Persian Gulf War Oil Spill (1991) and BP's Deepwater

DOI: 10.4324/9781003041627-27

Horizon Spill (2010), along with disasters such as the Exxon Valdez tanker disaster (1989).

For the oil industry, these events were part of the cost of doing business. But they were too frequent to be considered the exception to the rule. Oil spills could be large or small, and they could occur where oil is "drilled, stored, handled, refined, transported, and transferred." If they were truly just a part of doing business, environmentalists have argued, shouldn't we be giving more attention to why we continue to extract energy sources in this way? Immediate concerns on how to deal with the accidents when they happen have evolved into questioning the aggressive exploitation of fossil fuels wherever they are found. What the spills have taught us is that they are energy issues, but also wildlife and marine life issues, pollution issues, and water issues. All of the questions challenge the long-standing objective of economic growth and progress at any cost.

The piece on the Ogallala Aquifer presents a classic case of "The Tragedy of the Commons"—the tension between the rights and wishes of the individual and the good of the collective. The ceaseless use of the vast aquifer on the American High Plains demonstrated how farmers, ranchers, and others had strong incentives to take untold gallons of water, especially to serve the needs of irrigation and grazing, in what had been a traditionally arid and climatically hostile region. In consuming the bountiful water resource that was the aquifer, individuals began to deplete it, which not only had impact on their own ventures, but at the expense of everyone else and the environment itself.

Ironically, the exploitation of the Ogallala Aquifer transformed the High Plains in such a way that it was difficult to cease exploiting it, even as water supplies dwindled. To do so, in the worst case, was to return the thriving agricultural marvel to its arid past. The search for solutions to this dilemma is ongoing, made more complicated by the threats of climate change.

The episode on Mexican water management raises questions about the value of shifting from public control of water to privatization. The water privatization movement, beginning in the late-twentieth century, started with the premise that a privatized approach to water delivery offered a better, more cost-effective, and efficient means of getting potable water to customers with traditional public service, or with limited or no service at all. To critics, however, privatization meant placing control of water supplies in the hands of businesses only seeking profits, and thus treating water as a commodity rather than a public good.

As privatization of many services spread worldwide, especially beginning in the Ronald Reagan administration in the United States and the Margaret Thatcher government in the United Kingdom, questions arose as to whether cities—and nation-states for that matter—were giving up too much control—and revenue—in providing services to its people that were being condemned as drags on annual budgets. Also at stake was the degree to which local control of services eroded local authority in general, and the degree to which—in seeking profits—water companies would ignore maintaining service infrastructures (Melosi, 2011, 181–196).

The Mexican case is a curious one, not only because water privatization became a controversial topic, but was applied unevenly and with varying degrees of success. The convulsions in Mexico caused by the rise of neoliberalism—and resistance to it—powerfully affected decisions about the control of many services, water included.

That Mexico turned so heavily to consuming bottled water added an additional layer of complication because it seemed to suggest that citizens did not trust public or private management to protect the purity of its water supply. While the consumption of bottled water led to debates about squandering resources and the excessive and wasteful use of plastics, in Mexico it also raised questions about the general state of water supply systems there (Melosi, 2011, 192).

Notes

Knight, Clifford B. (1965) *Basic Concepts of Ecology* (New York: Macmillan).

Melosi, Martin V., ed. (2011) *Precious Commodity: Providing Water for America's Cities* (Pittsburgh, PA: University of Pittsburgh Press).

Scheffer, Victor B. (1991) *The Shaping of Environmentalism in America* (Seattle, WA: University of Washington Press).

Worster, Donald (1977) *Nature's Economy* (Cambridge, UK: Cambridge University Press).

20 Mexico's Ixtoc 1 Oil Spill

The search for new supplies of petroleum and natural gas inevitably led to increased interest in offshore wells a little before the start of the twentieth century, which also set off controversies about potential spills and blowouts. According to behavior ecologist and eco-toxicologist Joanna Burger, "Accidental spills can occur wherever oil is drilled, stored, handled, refined, transported, and transferred. These spills can be either massive and catastrophic or chronic. Few other environmental problems are as common or ubiquitous or have the potential for immediate environmental damage and long-range effects" (Burger, 2004, 965). Until recently, the most devastating marine oil spill in the Western Hemisphere occurred between June 1979 and March 1980 at the site of the Ixtoc 1 experimental well in the Bay of Campeche in the Gulf of Mexico. By most accounts, the Ixtoc 1 oil spill ranked globally as the second or third worst oil spill of all time. Much was learned from the experience, and much went unlearned.

Oil spills can threaten human health and wildlife. In marine environments, oil from a spill or well blowout floats, destroying plankton by absorbing sunlight. Birds, fish, and mammals can be killed or injured through direct contact or when food sources are destroyed. If not contained, an oil spill can impact the food chain all the way up to humans. Coastal areas can be drenched in oil and badly polluted. Impacts of spills, however, are complex depending on the type of spill, the weather, the nature of the location, and the geology of the shoreline.

Blowouts occur when intense pressure in a well causes the oil to burst out violently. According to some studies, an offshore drilling platform will average one to three major spills, 25 medium spills, and 2,000 small spills during its 20 years of production (Hebert, 2011, 1015–1018). Since 1918, hundreds of oil spills occurred in Galveston Bay. The blowout on the Ixtoc 1 platform was particularly serious because containing the vast release of crude proved to be confounding and took ten months to complete. The impacts of the blowout extended from the Gulf of Mexico itself to the beaches of eastern Mexico and Texas. Some important lessons were learned from the sad event, but offshore oil production continued, nevertheless. While hurricanes or cyclones may be the most dramatic and potentially most destructive events associated with water, marine oil spills produce their own misfortunes.

DOI: 10.4324/9781003041627-28

The first oil discovery in Mexico occurred at Tampico on the Gulf Coast in 1910. Major offshore oil wells were drilled in the Bay of Campeche (*Bahia de Campeche*) during the 1970s. By the early 1980s, it became the highest oil-producing region in Mexico. Beginning December 10, 1978, the Ixtoc 1 was being drilled by the SEDCO 135, a semi-submersible platform. It had been leased to Mexico's national petroleum company Petroleos Mexicanos (PEMEX).

SEDCO 135—all 3,527 tons of it—was built by the Ingalls Shipyard of Pascagoula, Mississippi, in 1965 under contracts from Southeastern Drilling Company, Inc. (SEDCO). Initially, it was put into service in the Gulf of Mexico, but after five years was chartered by foreign companies and towed to coastal West Africa for drilling operations there. In 1975, SEDCO 135 returned to the Americas and began drilling near Trinidad. In May 1978, what amounted to a third foreign charter was approved by the U.S. Department of Commerce for Perferaciones Marinas del Golfo (PEMARGO), which was a drilling company under contract to PEMEX (Myer, 1984, 3). Ixtoc 1 was meant to be a geological exploration well to a depth of 18,044 feet. By June 1, 1979, it had reached a depth of about 11,792 feet.

The Bay of Campeche in the Gulf of Mexico in southern Mexico, where the well was drilled, was bounded by the Yucatan Peninsula to the east, the Isthmus of Tehuantepec to the south, and by southern Veracruz to the west. The shallow bay of roughly 160 feet covered an area of approximately 6,000 square miles. The southwestern Gulf of Mexico itself is an important, dynamic, and complex subsystem. It is highly influenced by runoff from major rivers and nutrient flow and dissolved organics from coastal lagoons and estuaries within a very large marine ecosystem. The bay has great biodiversity but also features extensive energy resources. The latter offered important economic value to Mexico, a country seriously in need of successful economic ventures (Soto et al., 2014, 1).

On the early morning of June 3, the drill bit on the platform entered an area of soft sedimentary soil causing a break or a reduction in the weight of the bit. This break led to a fracture in the piping resulting in a complete loss in drilling mud circulation. Oil pressure began to increase in the well column. Three losses of mud circulation occurred in all, but after restoring circulation with more mud on the first two occasions, the supply on the platform was exhausted. The pressure continued to rise as the drill piping was being removed. A surge in mud pushed up the drill pipe and onto the platform. A blowout preventer stack was supposed to engage and shear through the drill pipe closing it together thereby trapping in oil prior to a potential blowout. But the preventer failed, and a blowout occurred.

Soon, escaping oil ignited and exploded when it made contact with fumes from a motor powering a derrick on the platform. The fire, which lasted until 10:00 a.m. the following day, caused the SEDCO 135 to collapse. The platform was towed northwest of the well site and ultimately scuttled (Ixtoc Oil Spill; Myer, 1984, 8, 11–12). Oil rapidly flowed out of the well following the blowout with anywhere from 30,000 to 45,000 barrels a day. For a time, oil was gushing out of the well 20 feet into the air. (No injuries or death were initially reported.)

This kind of accident at Ixtoc 1 was atypical and somewhat rare, although not unprecedented. The Santa Barbara well #21 oil spill (California) in 1969, the Laban Island oil spill (Iran) in 1971, and the Ekofisk Bravo oil rig blowout in the Norwegian Sea in 1977 were of a similar type. In each case, oil eruptions stopped after 10 days. In the case of Ixtoc 1, the well took 10 months to cap. It is unclear what the total quantity of the spill in the Bay of Campeche amounted to; estimates range from about 476,000 tons (about 140 million gallons) to 1,653,450 tons. One-third to one-half of that amount burned in the air, creating significant air pollution (Ixtoc 1, 1999).

Maybe because of the unique character of the blowout or lack of adequate preparation, stopping the flow proved to be a monumental task: "Oil gushed into the Gulf of Mexico at a staggering rate from the damaged riser that had attached the platform to the well. Nobody knew what to do, although engineers tried various measures to stem the flow, including a containment dome [also skimmers and floating barriers]. Chemical dispersants to break up the oil were applied at one of the highest rates in history. Some of the oil was trapped well below the Gulf's surface, with undetermined effects. It seemed as though the spill might drag on forever" (Schrope, 2010, 304).

Figure 20.1 Ixtoc 1 oil well blowout, June 1979 to March 1980. Public domain. Wikimedia.

PEMEX hired blowout control experts and other spill-control experts including the famous Red Adair from Houston. The blowout preventer (BOP), consisting of valves on the well pipe over the seafloor, had not been damaged by the blowout and sat above the well casing where the leak occurred. Adair attempted to clear the blowout preventer shears and reclose them which reduced the flow but only for a few hours. Martec International of Houston had 50 workers on site, a remotely operated vehicle called TREC, and the submersible Pioneer 1. The TREC tried to find an effective route to the blowout preventer, but the approach was made difficult by poor visibility and large amounts of debris in the water. PEMEX also hired Mexican diving Company Daivaz, who sent divers (some of whom were killed in the process) to activate the BOP, but the pressure of oil and gas caused the valves to rupture. (The BOP was reopened to prevent it from being destroyed.)

None of these efforts completely stopped the flow of oil, but gradually the amount of leakage began to subside. The rate of release was reduced by one-third in early August by pumping about 100,000 metal balls into the well. In the months that followed Mexican authorities drilled two relief wells to reduce stress in the main wellhead which lowered the pressure further. While attempts to find a way to cap the well were underway, Norwegian experts were under contract to bring skimming equipment and containment booms to the site to begin the cleanup.

On March 23, 1980, the relief wells succeeded in stopping the flow by allowing the pumping of mud into the main wellhead. This led to sealing the well with several cement plugs. Experts were divided in evaluating PEMEX's response to the disaster. Some praised the company for exhausting available technical solutions; several others considered the methods employed unproductive. The consequences of the spill now became the big story.

Oil covered some Mexican beaches almost a foot thick at a time when Kemp Ridley turtles were hatching at Rancho Nuevo. Oil also affected about 160 miles of U.S. beaches. Along the shoreline in south Texas approximately 3,900 tons of oil-covered beaches from the middle of August to early September 1979. However, most of the beach oil there disappeared when a tropical depression passed through the area. Mixed sand/shell beaches suffered more damage than beaches with fine-grained sand. But more expansively shrimp nurseries, mangroves, and seabirds were covered in oil. The amount of oil reaching the bottom sediments of the Gulf was estimated at more than 130,000 tons.

In the long run, marine life suffered more than the beaches. The spill especially damaged larvae and juvenile shrimp which were highly sensitive to oil contamination. Zooplankton, vital in the marine food chain, decreased almost fourfold between 1979 and 1982 in the Southern Gulf of Mexico. About 10,000 baby Ridley turtles were cleaned and let loose, but there was no way of determining what the long-term impact on the turtles might be. Around Isla Arena in Campeche oysters disappeared.

From an economic standpoint, the fishing and shrimping industry sustained a major blow. Fish catches took between three and five years to improve; shrimpers did not see normal hauls for about two years. Recreational enclaves along the Mexican and U.S. coastline experienced serious economic impact with a major loss of tourism.

PEMEX spent about $100 million (more than three times that amount today) to clean up the spill but avoided a large share of compensation costs by asserting sovereign immunity as a state-run business. An estimate of the total cost of the spill was $1.5 billion, of which $0.4 billion was for response expenses and $1.1 billion for damages. Observable damage after the spill allowed for only anecdotal evidence of long-term environmental harm to the Gulf. Neither the Mexican nor American government produced scientific or financial reports related to the accident. There was little study devoted to fish populations. Funding for studies of the impact of the spill disappeared quickly after the event, although a few groups attempted to determine the extent of the damages—at least superficially. A combined research cruise by the National Oceanic and Atmospheric Administration (NOAA) ship *Researcher* and a contract vessel *Pierce* collected more than 1,000 samples for biochemical and microbiological study in September 1979 and completed detailed hydrographic surveys around Ixtoc 1 that helped advance information on how oil behaved in the Gulf.

The lack of comprehensive data and analysis about the Ixtoc 1 spill did not shield the fact that ocean drilling and greater oil tanker traffic guaranteed more blowouts and spills in the future. Indeed, the Ixtoc 1 blowout renewed a controversy over oil exploration along the continental shelf that went back at least to the 1930s in the United States. During the New Deal Interior Secretary Harold Ickes questioned state ownership of submerged coastal lands. In 1945, President Harry Truman issued an executive order proclaiming that the continental shelf was under federal control, which clearly angered the oil industry and its supporters. Finally, in 1953 President Dwight Eisenhower signed the Submerged Lands Act. It granted California, Florida, Texas, and Louisiana rights to lease submerged areas within three nautical miles. (With court decisions in their favor, Texas and Louisiana were able to extend their jurisdiction to about 10.5 miles.) Eisenhower also signed the Outer Continental Shelf Act in that same year which gave the federal government exclusive jurisdiction over the ocean bottom beyond the newly established state limits (Federal limits were established by international law.). Mexico set its federal limits also in accordance with international law (about 200 nautical miles from the country's shoreline).

Nothing riveted attention on the environmental dangers of oil production like the Santa Barbara oil spill on a federal lease in January 1969 when Union Oil's Well A-21 blew. The leak released 235,000 gallons of crude with a slick of 800 miles. The blowout raised questions about exploiting offshore oil, the aesthetics of oil rigs along recreational coastlines, corporate responsibility for environmental disasters, and the need for environmental protection. With the much larger blowout at Ixtoc 1, questions of international jurisdiction and liability further complicated the debate.

In the late 1990s, production in deepwater surpassed that in shallow water for the first time. In 2010, a blowout of BP's Deepwater Horizon in the Gulf of Mexico became the largest marine oil spill in the history of the petroleum industry. An estimated 2.6 million gallons a day from the leak for 87 days before it was sealed. Pulitzer prize-winning historian Jack E. Davis argued, "Since the *Deepwater Horizon* tragedy of 2010, oil has hijacked the Gulf's identity. It frames how we—from journalists to policymakers, even scientists and tourists—perceive the American Sea" (Davis, 2017, 10). While the death toll was greater than Ixtoc 1 (11 people), the massive oil spill from both wells was mind boggling and the cleanup efforts roundly criticized.

The spills invoked outrage, but no moratorium of offshore drilling shut down the process. The debate between continuing oil and gas needs and wants and the environmental implications of such events remains unresolved. The proverbial saying that oil and water don't mix may still apply, but oil spilling in oceanic waters continues to be a fact of modern life.

Notes and Further Reading

Burger, Joanna (1997) *Oil Spills* (New Brunswick, NJ: Rutgers University Press).

Burger, Joanna (2004) "Oil Spills," in Shepard Krech III, J.R. McNeill, and Carolyn Merchant, eds., *Encyclopedia of World Environmental History*, vol. 3 (New York: Routledge).

Davis, Jack E. (2017) *The Gulf: The Making of an American Sea* (New York: Liveright Pub. Corp.).

Gundlach, Erich R., et al. (1981) "Impact and Persistence of Ixtoc 1 Oil on the South Texas Coast," *International Oil Spill Conference Proceedings*: 477–485.

Hebert, Kimberly (2011) "Oil Spills," in Kathleen A. Brosnan, ed., *Encyclopedia of American Environmental History*, vol. III (New York: Facts on File, Inc.).

"Ixtoc Oil Spill," online, https://sites.google.com/a/asu.edu/ixtoc-oil-spill/home.

"Ixtoc 1" (1999) *Cedre*, February 1, online, https://wwz.cedre.fr/en/Resources/Spills/Spills/Ixtoc-1.

Jernelov, Arne and Olof Linden (1981) "Ixtoc 1: A Case study of the World's Largest Oil Spill," *Ambio* 10: 299–306.

Melosi, Martin V. (1985) *Coping with Abundance: Energy and Environment in Industrial America* (New York: Knopf).

Myer, Peter G. (1984) *Ixtoc 1: Case Study of a Major Oil Spill* (Thesis, Marine Affairs, University of Rhode Island, Spring).

Schrope, Mark (2010) "The Lost Legacy of the Last Great Oil Spill," *Nature* 466 (July 14): 304–305.

Soto, Luis A., et al. (2014) "The Environmental Legacy of the Ixtoc-1 Oil Spill in Campeche Sound, Southwestern Gulf of Mexico," *Frontiers in Marine Science* 1 (November): 1–9.

21 The Ogallala Aquifer in Decline

Previous case studies in this volume have considered irrigation practices, especially among the Hohokam and Spanish settlers in New Mexico. Others, like the essay on placer and hydraulic mining in the California Gold Rush, describe the exploitation of freshwater supplies and the dire implications of doing so. The story of the Ogallala Aquifer uniquely connects the emergence of a major irrigated landscape in the heart of the United States and its catastrophic depletion in recent years. The Ogallala transformed a vast semiarid region into a fertile agricultural Eden and its overuse is threatening to return the region to a bleak past.

The Ogallala—or High Plains—Aquifer is one of the world's largest aquifers (and one of its largest bodies of fresh water) at one point holding about 3 billion acre feet (or 1 quadrillion gallons) of water spread over an area of 174,000 square miles encompassing eight states. These states run about 800 miles from South Dakota to Texas and New Mexico (including South Dakota, Nebraska, Wyoming Colorado, Kansas, Oklahoma, New Mexico, and Texas).

One estimate suggests that the volume of water in the aquifer would cover all 50 states in more than 1 or 1/12 feet of water. The Ogallala contained more water than the Mississippi River had carried to the Gulf of Mexico in 200 years. Some call it America's 6th Great Lake. If drained it would take more than 6,000 years to refill it naturally. Two-thirds of the water volume in the Ogallala lies under Nebraska (with only 37 percent of the area). In some places there, it is more than 1,000 feet thick, although the average depth throughout the whole aquifer is about 500 feet.

According to the U.S. Geological Survey (USGS), an aquifer is "a geologic formation, a group of formations, or a part of a formation that contains sufficient saturated permeable material to yield significant quantities of water to wells and springs" (USGS, 2021). The rock contains water-filled pore spaces (or voids). The Ogallala is a "confined aquifer," that is, it is overlain by a rock layer that does not transmit appreciable amounts of water (thus impermeable). The water, therefore, must be removed by a well or another means of tapping it. The water found in aquifers is replenished by drainage (a groundwater recharge) which is a very slow process.

DOI: 10.4324/9781003041627-29

The Ogallala was formed by runoff from the Rocky Mountains that became trapped under the modern Great Plains. Sediments and rocks in the formation range from 33 million years old, but most are less than 12 million years old. The aquifer was first discovered in the 1890s. It was named by geologist N.H. Darton in 1898 in reference to the town nearest to where the rock formation associated with the aquifer was identified—Ogallala, Nebraska. As one writer noted, "As is often the case in US History, colonial-settler communities adopted this city name from the people whose land was stolen [Oglala Nation]..." (Undlin, 2021).

At this point, the water remained largely inaccessible for several years. The aquifer was first tapped for irrigation in 1911, but in very small amounts. Droughts and especially the Dust Bowl in the 1930s prompted more attention to the aquifer. It was not until after World War II that improved pumping and irrigation technology made it available for large-scale use.

Not only the size of the aquifer, but its location made its impact on American history momentous, especially in the development of agriculture. Environmental author William Ashworth observed:

> If you snack on popcorn or peanuts, you are probably eating Ogallala water; if you dress in cotton clothing, you are probably wearing it. The Ogallala grows wheat and milo, sunflowers and sorghum. It grows alfalfa for cattle, and it grows the cattle as well...
>
> (Ashworth, 2006, 10).

Environmental historian John Opie explained, above all else "agriculture is climate dependent to the extreme." He added, "Aside from Alaska, nowhere in the United States was this agricultural vulnerability to a climate more formidable than on the High Plains (also known as the Great Plains)" (Opie, 1998, 355). The High Plains—making up about one-sixth of the contiguous U.S.—runs north-south through the United States and is approximately 300 miles wide, and is drained eastward by the Platte, Arkansas, and Canadian rivers. It essentially bisects the country with the 98th meridian as the eastern border and the Rocky Mountains on the west. The altitude of the plains can reach 5,000 feet above sea level. The climate there can be measured in extremes—hot, dry, and windy in the summer, and severely cold in the winter. It is semiarid, with rainfall averaging 12–20 inches per year. (It is considered semiarid because most farming in the United States has taken place where rainfall averaged 30–40 inches per year.)

The High Plains initially was covered with a variety of grasses. There were low hills, shallow valleys, and small steams. Again, according to Opie, "The balance involved poorly understood natural forces, such as enormous grass fires, climate fluctuations, animal grazing, together with low-level Native American interference" (Opie, 1998, 356).

In the nineteenth century, people such as Zebulon Pike influenced thinking about the High Plains. Early in the decade, the moniker of the

"Great American Desert" was common among travelers and explorers in the region and found its way on maps, suggesting that the area was worthless for farming. Such an image was powerful, especially at first glance at a treeless grassland that stretched for miles in all directions and seemed absolutely barren.

Such a view was reinforced by white settlers who characterized all the various Plains Indian tribes—from a Sarcee to the north (in present-day Canada) to the Tonkawa to the south (in central Texas)—as nothing but nomadic hunters of buffalo with limited connection to the land. Not until the 1500s, with the introduction of the horse by the Spanish, that tribes who acquired horses were dependent on farming (Waldman, 2006, 225–226). The assumption of many whites was that Plains Indians eschewed farming because they did not know how to do it or that they cared little about land ownership or control. Such stereotypes were simple justification for undervaluing Native American culture or eliminating the tribes themselves.

White homesteading in the High Plains began in the 1870s and continued into the 1920s. The urge to acquire land often camouflaged the stark realities facing settlers who hoped to farm there. Many settlers bypassed the plains in hopes of finding fertile land to the West. Others were not deterred by predictions that the lack of water and wood made the plains unsuitable for farming, but possibly only cattle grazing at best. They also were hampered because they were accustomed to farming conditions in the eastern United States characterized by moister soil. The myth that "rain followed the plow" was deeply rooted in the thinking of those eager for their own plot of land.

The Homestead Act (1862) that limited settlers to 160 acres of land was unworkable for promoting farming on the semiarid grasslands. Frigid winters, drought, and grasshoppers forced farmers to shift from growing corn to growing winter wheat with techniques not suited to the unfamiliar soil conditions or untested in America. A variety of new federal laws, which sought, among other things, to aid the settlers on the High Plains, failed. For example, the Timber Culture Act (1873) gave free title of an additional 160 acres to any farmer who planted trees on one-fourth of the plot. The idea of improving the climate on the plains by this method failed miserably. The government through subsidies and supportive programs also urged the use of irrigation, but where would the water come from?

From 1878 to 1887, unlikely heavy rains came to the High Plains and gave hope that farming bonanzas were possible. But drought returned in summer 1887, and more drought followed in subsequent years. In the 1920s, grain prices drastically fell, and then the Dust Bowl struck in the 1930s.

In the High Plains environment, accessing sufficient and consistent water demanded looking beyond the surface or wishing for heavy rains to return. But not until the 1930s, did geologists with new drilling rigs learn about the vast groundwater supplies that existed below the plains. They would be shocked to discover how much water was there. According to Dr. John Tracy, director of the Texas Water Resources Institute, "Technological innovation, financial and economic conditions, infrastructure changes, social values—all these factors drive change" (AgriLife Today, 2021). In this case, new technology was the spearhead to tapping the aquifer.

Early settlers dug wells by hand on their homesteads, but these wells were only good enough to supply immediate needs. By the 1880s, farmers erected thousands of windmills to pump water out of the ground. The most successful ones incorporated a vane to automatically turn the blades into the wind. Some farmers constructed small earthen reservoirs to store water for modest irrigation. These windmills could only reach about 30 feet, and the reservoirs could flood only about 10 acres of land (they needed about 16 times that amount).

It was irrigation, not dry farming methods and planting drought-resistant crops, that seemed to promise the greatest opportunity to subdue the plains. Traditional irrigation, which demanded gravity diversion from streams, rivers, and reservoirs, could work in some parts of the western United States, but the paucity of running surface water or limited construction of dams and reservoirs made such an approach difficult or downright impossible on the Great Plains. Some farmers acquired land close to major rivers or streams, but they often were faced with short and/or unreliable supplies of water. Tapping the aquifer virtually eliminated surface-water irrigation (Rogers, 2013, 4–5).

By the turn of the century, the beginning of commercial irrigation was made possible by utilizing more elaborate machines. New centrifugal pumps provided larger volumes of water per minute. By 1900, there were horizontal and vertical centrifugal pumps. The former was used in shallow wells; the latter in deeper wells below 25 feet. They both were powered by an electric motor or an internal combustion or steam engine. These early pumps, however, were not suitable for exploiting the Ogallala. The newer "pitless" pumps that could be set in deep-drilled wells offered a better approach. In addition, because of the early stage of electrical power generation, the internal-combustion engine seemed to be the cheapest available power source by 1910.

There was a great deal of enthusiasm for the emerging pumping technology, but the cost often was out of reach for many farmers. Mechanical problems also plagued the equipment. Markets for irrigated crops and absence of credit to finance pumping plants were immediate setbacks for the farmers, as were good rainfall amounts in the 1920s. By the 1930s, new plants proved more efficient (deep-well turbine pumps), had better power sources (cheap natural gas), and had broader applications, such as in municipal water-supply systems (Kromm and White, 1992, 28–39). New technology and cheap fuel made modern irrigation possible.

For irrigation, improved water delivery systems were essential. The main form of surface irrigation was furrows, which were plowed between crop rows where water could flow. Irrigation was easier by using light-weight aluminum pipe for running water into the rows, but underground concrete pipelines and later plastic underground pipelines also were used. Sprinkler irrigation—which conserves more water—became especially widespread with the adoption of center-pivot irrigation. Initially patented in 1949 by

Frank Zybach, who used the system for sugar beets; a second patent was issued in 1952 to Zybach and A.E. Trowbridge. They sold the rights to produce and sell the system to Valmont Industries of Valley, Nebraska who marketed it as "Valley" systems. Soon other iterations appeared on the market (Kromm and White, 1992, 40–41).

Center-pivot irrigators are lateral pipes with spray nozzles usually suspended on drop tubes and mounted on wheeled structures (or towers). They were connected by swivel to a centralized well or to underground pipe. The power was normally supplied through diesel engines. The center pivot is anchored at the center of a field and rotates in a circle with a typical radius of one-quarter mile. By 1979, there were more than 15,000 units in Nebraska alone, and as eminent environmental historian Donald Worster stated, center-pivot systems "transformed the plains landscape from a giant checkerboard to rows and rows of bright green checkers" by opening up fragile land to cropping (Worster, 1985, 313).

Optimism for tapping the agricultural possibilities of the High Plains now exploded with the advent of the new mechanized pumping and sprinkler irrigating systems. Historian Donald E. Green observed, that while the changes in exploiting the land brought prosperity to many, "...dependence on the new systems has also created a kind of secular faith that technology is infinitely capable of solving future problems involving overdraft of the Ogallala and pollution of the aquifer caused by chemical fertilizers and pesticides" (Kromm and White, 1992, 42).

That optimism seemed to be warranted as farmers and others began aggressively to draw water from the Ogallala beginning in the 1940s. In West Texas, for example, the number of irrigation wells rose from 139 in 1914, to 1,166 in 1937, to 27,983 in 1954, and then to 66,144 in 1971. In the region, water was being pumped by nearly 200,000 irrigation wells since the 1940s. Nebraska irrigated fewer than 1 million acres in 1959 to about 7 million in 1977—mostly using the Ogallala.

By the turn of the twentieth century, the Ogallala was providing groundwater for about 20–27 percent of all irrigated land in the United States, 90 percent of which was used for farming. This translated into water withdrawals of about 16.6 million acres-feet of water annually (Verchick, 1999, 20–21; Reisner, 1987, 437; Guru and Horne, 2001, 321–322). In 1950, plains farmers irrigated about 3.5 million acres of farmland; in 1990 that number was about 15 million.

Not surprisingly, the land above the aquifer came to be regarded as "America's Breadbasket" producing great amounts of corn, alfalfa, soybeans, wheat, and also cotton. The aquifer supported one-sixth of the world's grain. The Ogallala also provided water for millions of beef cattle and other livestock. Nebraska (6.3 million) and then Texas (4.4 million) lead the Great Plains states in irrigated acreage—mostly from Ogallala water. The High Plains region also relied on the available groundwater for residential and industrial uses, including 82 percent of the area's drinking water.

The great boon in finding what appeared to be an endless supply of fresh water to remake the High Plains into highly productive farmland blinded people to the reality that groundwater use had limits. Ashworth stated this point well: "The thread that runs through all of these changes [in farming the High Plains] is an assumption that the Ogallala Aquifer will survive—at least for a while, at least in some form. All the new patterns require access to more water than plains rains can dependably supply. The Ogallala is the only other available water source. For the most part, High Plains residents understand this. But there is a curious disconnect between what they understand and what they do…" (Ashworth, 2006, 58).

The book, *Ogallala: Water for a Dry Land*, explains the stark reality of the aquifer:

> Unlike most of the world's water supplies, Ogallala groundwater is not renewable. It is "fossil water," drawn over twenty-five thousand years ago from the glacial-laden Rockies before being geologically cut off, perhaps diverted away by the Pecos and Rio Grande Rivers. Groundwater replacement trickles down from the surface at a rate of only an inch a year, while pumping is measured in feet a year. Nothing can accelerate that trickle, and artificial replacement remains impossible.
>
> (Opie et al., 2018, 1)

Too many people failed to understand that the water in the aquifer did not flow like a river but was trapped and essentially finite.

The unsustainable farming practices of the Dust Bowl era in the High Plains were replaced with new irrigation technology—promoted by various government programs and subsidies—which have proven unsustainable in their own right because of the heavy draw on the Ogallala.

A recent study done at Kansas State University predicted that the aquifer would be 70 percent depleted by 2060 if irrigation practices remained the same; if farmers cut their use by 20 percent the aquifer could last another 100 years (Frankel, 2018). One study stated that more than half of currently irrigated land in portions of Western Texas, Eastern New Mexico, and the Oklahoma Panhandle could be lost by the end of the twenty-first century; 80 percent of that amount by 2060. Other areas near the central part of the aquifer are also vulnerable (*Houston Chronicle*, September 12, 2021, A9).

In 1999, environmental law professor Robert R.M. Verchick predicted that the aquifer would be gone in 20–30 years. "We are talking about a resource with a shorter life span than Social Security" (Verchick, 1999, 13). Marc Reisner writing back in the late 1980s projected that if energy prices rise, farm acreage shrinks, and commodity prices fail to give farmers adequate returns, "The odds are high…that long before all the water runs out, the farmers will no longer be able to afford to pump" (Reisner, 1987, 438). There was the possibility that the rate of pumping water out of the aquifer could improve with better irrigation techniques and technology, but new equipment also costs more money.

Such water depletion is not uniform. For example, areas in the Texas Panhandle and in Kansas have seen their water supplies seriously diminished while other areas have not yet experienced such an impact. In 2019, farmers and ranchers in Eastern New Mexico reported that five of its seven center-pivot irrigation systems could not pump water, while farmers and ranchers in north-central Nebraska had not noticed an appreciable decline in the water table.

Depletion and lack of recharge were serious long-term issues and de-manded immediate attention. Water pollution issues of various types only complicated the situation. Especially in recharge areas (like the Nebraska Sandhills), agricultural chemicals used for fertilizing—such as nitrates—affect groundwater quality, as do pesticides and animal manure from feedlots. While the quality of aquifer water generally has been suitable for irrigation, in several locations the water does not meet U.S. Environmental Protection Agency (EPA) drinking water standards (e.g., dissolved solids, salinity, fluoride, chloride, and sulfate) (Guru and Horne, 2001, 322).

The pollution of groundwater caught many people by surprise. While the depletion of groundwater was a concern, the question of quality of the water had rarely been an issue. There was an abiding faith that groundwater is in-trinsically purer and freer from pollutants than surface water. Beginning in the 1960s and 1970s, that hallowed belief came into question, but it has taken time to address groundwater pollution and suggest solutions.

State and federal protection of groundwater could be identified in bits and pieces of legislation in a variety of laws enacted in the 1960s and later. Going back to the nineteenth century, laws dealing with groundwater focused on rights and remediation rather than groundwater quality. The Clean Water Act (1972)—the basic structure for regulating pollution in water—deals more broadly with surface water than groundwater. In the 1980s, Congress and some federal agencies made efforts to bolster groundwater protection au-thority. In 1984, the EPA established the Office of Ground Water Protection and released its groundwater-protection strategy. This was not enough to appease those concerned with the problem. The following year, the Conservation Foundation and the National Governors' Association organized a National Groundwater Policy forum that produced recommendations for a national groundwater-protection plan. Some legislation dealing with ha-zardous and toxic wastes applies to groundwater. Yet, a comprehensive federal groundwater law awaits the future (Melosi, 2000, 385–386, 390–391).

Pollution concerns also intensified with the announcement of the proposed Keystone XL pipeline in 2008. The pipeline was intended to transport Canadian tar sands crude from Alberta to various processing sites in the United States with about 830,000 barrels of crude per day going to refineries on the Gulf Coast of Texas. Those opposing the pipeline feared that it would threaten the Ogallala water and Native American lands. The controversy inspired intense protest and political rancor. However, on January 25, 2021, newly elected president Joseph Biden shut down the project by executive order.

Aside from pollutants, overdrafting can lead to subsidence, where the land surface cracks or drops as the water is depleted from the aquifer. Subsidence can damage a variety of surface structures and increase flooding. Underground, the settling of the soil can reduce the storage capacity of the aquifer. Since recharge is so slow in the Ogallala, this settling might not appear to be too severe in the short run (Glennon, 2002, 33–34).

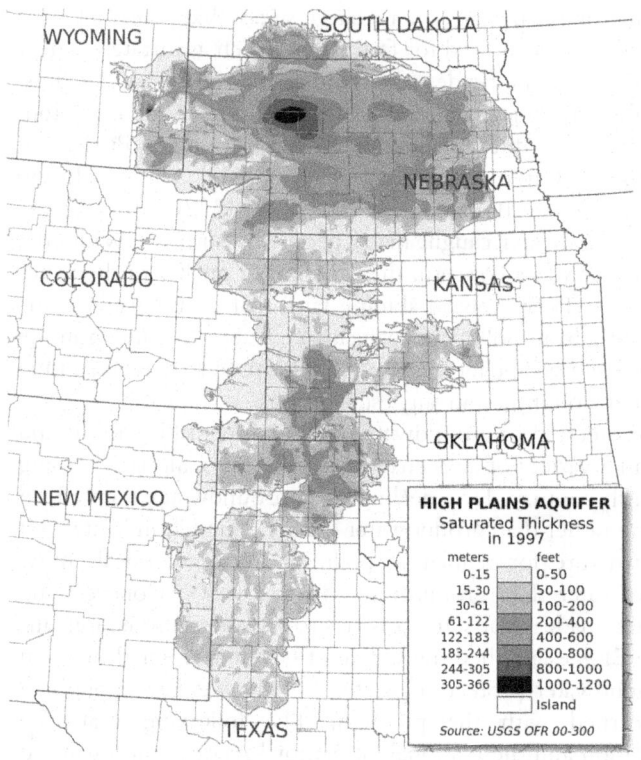

Figure 21.1 Digital map of the saturated thickness of the High Plains aquifer in parts of Kansas, Nebraska, New Mexico, Oklahoma, South Dakota, Texas, and Wyoming, 1996–1997. From GIS data produced by the USGS and published in Open File Report 00–300 (USGS OFR 00–300)[1]. Creative Commons Attribution-Share Alike 3.0. Wikimedia.

Finding answers to the array of problems facing the Ogallala is complicated by the numerous players involved, especially the eight states sitting atop the aquifer. As Verchick stated, "The problem, of course, is that the states all have different interests and their independent legal systems encourage officials to pretend as if the Ogallala were bounded by their state borders." As of 1999, New Mexico followed an "unlimited use" strategy to support economic development. Oklahoma operated on the idea of "planned depletion in 50 years." While Kansas supported "zero depletion," that is, a system where consumption would be reduced to match the recharge rate (Verchick, 1999, 21–22).

Over the last several years, some states achieved some success in slowing depletion, such as Kansas and Nebraska, but not others. Interstate compacts have been struck in recent years in recognition of regional concerns. Kansas, Nebraska, and Colorado apportion "virgin water" from the Republican River Basin that feeds into the Republican River. The federal government, for its part, instituted the Ogallala Aquifer Initiative to provide technical and financial support for conservation practices (Frankel, 2018). But the sad fact is that much of the High Plains is "locked into high water consumption to grow the wheat and water the beef" (Guru and Horne, 2001, 325).

Other than developing strict limits on water use, some farmers have taken to shifting to crops that require less water or utilizing more efficient irrigation systems. Some are turning to plant native grasses that require little rain and grow dense roots to hold the soil in place. "There's a reason Mother Nature selected those plants to be in those areas," Nick Bamert of Muleshoe, Texas stated. "The natives...will persist because they've seen the coldest winters and the hottest dry summers."

Old-style programs such as crop insurance and land conservation programs did little to address the depletion problem. In some cases, farmers planted crops even if they might fail because they were covered by insurance. And increasing yields were often more enticing than taking government payments for restoring grassland. From 2016 into 2021, less than 328,000 acres of land were involved in the Department of Agriculture's Grassland Conservation Reserve Program (*Houston Chronicle*, September 12, 2021).

Depletion and other issues are not exclusive or limited to the Ogallala Aquifer. You need look no farther than California's Central Valley or as far away as India and China. Such problems not only suggest past or current local agricultural conditions but the longer-range future impacts of climate change. On the Great Plains, warmer temperatures—including in the winter months—are becoming more typical. The average temperature in North Dakota, for example, has increased faster than any other state in the lower 48. Such rising temperatures lead to greater irrigation demands, potentially less agricultural production, and reduced variability in crops that can be grown. Ultimately, climate change may cause a northward shift in lands used for farming (Lennon, 2020). Some predictions suggest that climate change could make drought longer lasting in the Ogallala region and possibly more intense in the next 50 years.

Ogallala: Water for a Dry Land aptly ends with the discussion of "The Tragedy of the Ogallala Commons." The idea of "the tragedy of the commons" was first used by biologist Garrett Hardin in 1968 to explain how those with open access to public resources act independently and, in their self-interest, to deplete those resources contrary to the common good. *Ogallala* suggests that as the states in the High Plains realized that decline of the aquifer was real, they attempted to devise ways to limit consumption of water:

> They have pursued a variety of different paths to conserve the Ogallala while still consuming it because there was a broad realization that there were no substitutes for Ogallala groundwater. Yet all these strategies were sorely tested by recurrent droughts intensified by climate change. The Ogallala declines in dry years and in wet years, with few exceptions.
>
> (Opie et al., 2018, 342)

Learning from the past has been difficult for the people of the High Plains. They experienced desolation followed by great successes, followed by uncertainty. History offers some hints on how to deal with the present and the future but remaking a culture and society is never easy. This is in many ways a classic case of path dependence when the present state of a system is constrained by its history, where past choices limit future choices—but hopefully don't prevent them.

Notes and Further Reading

Ashworth, William (2006) *Ogallala Blue: Water and Life on the High Plains* (Woodstock, VT: The Countryman Press).

Frankel, Jeremy (2018) "Crisis on the High Plains: The Loss of America's Largest Aquifer—the Ogallala," *University of Denver Water Law Review*, May 17, online, http://duwaterlawreview.com/crisis-on-the-high-plains-the-loss-of-americas-largest-aquifer-the-ogallala/.

Glennon, Robert (2002) *Water Follies: Groundwater Pumping and the Fate of America's Fresh Waters* (Washington, D.C.: Island Press).

Guru, M. and J. Horne (2001) "The Ogallala Aquifer," *Transaction on Ecology and the Environment* 48: 321–328.

Hornbeck, Richard and Pinar Keskin (2014) "The Historically Evolving Impact of the Ogallala Aquifer: Agricultural Adaptation to Groundwater and Drought," *American Economic Journal: Applied Economics* 6: 190–219.

Kromm, David E. and Stephen E. White, eds. (1992) *Groundwater Exploitation in the High Plains* (Lawrence, KS: University Press of Kansas).

Lennon, Dan (2020) "Climate Change and the Ogallala Aquifer," *E: The Environmental Magazine*, October 6, online, https://emagazine.com/climate-change-and-the-ogallala-aquifer/.

Melosi, Martin V. (2000) *The Sanitary City: Urban Infrastructure in America from Colonial Times to the Present* (Baltimore: Johns Hopkins University Press).

"Ogallala Aquifer Depletion: Situation to Manage, Not Problem to Solve" (2021) *AgriLife Today*, March 19, online, https://agrilifetoday.tamu.edu/2021/03/19/ogallala-aquifer-situation-to-manage-not-problem-to-solve/.

Opie, John (1998) *Nature's Nation; An Environmental History of the United States* (New York: Harcourt Brace College Publishers).

Opie, John, Char Miller, and Kenna Lang Archer (2018) *Ogallala: Water for a Dry Land* (Lincoln, NEB: University of Nebraska Press, 3rd ed.).

Reisner, Marc (1987) *Cadillac Desert: The American West and Its Disappearing Water* (New York: Penguin Books).

Rogers, Jedediah S. (2013) *The High Plains Groundwater Demonstration Program* (Historic Reclamation Projects, Bureau of Reclamation, July), online, https://documents.pub/document/high-plains-states-groundwater-recharge-demonstration-program-.html

Undlin, Siri (2021) "Why Is the Ogallala Aquifer So Important?" *Earth.com*, online, https://www.earth.com/earthpedia-articles/why-is-the-ogallala-aquifer-so-important/.

USGS (2021) "Principal Aquifers of the United States," online, https://www.usgs.gov/mission-areas/water-resources/science/principal-aquifers-united-states?qt-science_center_objects=0#qt-science_center_objects (accessed September 21, 2021).

Verchick, Robert R. M. (1999) "Dust Bowl Blues: Saving and Sharing the Ogallala Aquifer," *Journal of Environmental Law and Litigation* 14: 13–23.

Waldman, Carl, ed. (2006) *Encyclopedia of Native American Tribes* (New York: Facts on File, 3rd ed.)

Worster, Donald (1985) *Rivers of Empire: Water, Aridity, and the Growth of the American West* (New York: Pantheon).

22 Water Management and Privatization in Modern Mexico

According to journalist Carmelo Ruiz Marrero, "One of the conclusions reached at the [First People's Workshop in Defense of Water in Mexico City in 2005] was that Mexico is a beachhead for privatization throughout the region" (Marrero, 2005). An indicator of changing water consumption in Mexico, according to one source, is that in 2015 it was leading the world in the consumption of bottled water. The rapid growth of that business has opened a controversy over the energy-intensive and environmentally unfriendly practice that water in plastic bottles garners worldwide.

This essay deals with water management in Mexico, and the degree to which international efforts at privatizing water-supply services is where Mexico is heading. Critics suggest that water corporatism undermines public control of its water, replacing it with handling of that vital resource by private transnational water companies. To some, this is a question of the human right to water.

At the April 2005 meeting of the above-mentioned workshop more than 400 people of all walks of life from Mexico and throughout the Western Hemisphere met to discuss water issues. The intent of the meeting was to share experiences about privatized water services and the attempts of governments to transfer water management to transnational companies. In the numerous presentations, participants discussed reactions to these services being promoted in their communities.

At the top of the agenda was the consideration of defending access to clean water as a human right to be managed in a responsible way. Participants from countries such as Nicaragua, Bolivia, and Ecuador who had fought against water privatization spoke about what they believed were questionable justifications for abandoning a public option and replacing it with management in the private sector.

Yet, the path to privatization of water in Mexico has not been uniform and hardly absolute. The country faced a crisis in the early twenty-first century with 2 percent of the global population and only 1 percent of the global water supply. Steady demand for more water (in cities and in the countryside) increased pressure on all levels of government to seek new paths in water management and delivery. But as resource specialists Christopher A. Scott and

DOI: 10.4324/9781003041627-30

Jeff M. Banister argued, "There are two separate and incongruous water management systems in place in Mexico. One, the official system, is derived from decades of centralized water and financial resource allocation that is firmly rooted in Mexico City. The second, a nascent form of decentralized autonomy within official institutions coupled with growing civil society demands and increasing public participation, is at loggerheads with the first" (Scott and Banister, 2008, 71).

Clearly evident on a world scale beginning in the 1970s, tension between viewing water in moral or legal-rights terms clashed with the practice of treating water simply as a resource (or commodity) like any other in a market economy. While the debate between water as a right (or public good) and water as a commodity has not been resolved, the acquisition of fresh water became a significant medium for dispute within organizational and environmental justice contexts. A serious question was whether fresh water supplies could be best provided and managed by public or private means. Going back to the early twentieth century, Mexico faced this dilemma that other countries confronted in earlier times or would confront in the future.

Since 1917, the Mexican Federal government maintained ownership of all surface and groundwater in the country and centrally managed those water resources. Water rights were a major theme during the Mexican Revolution. Article 27 of the revolutionary 1917 Mexican Constitution established the principle of "original nation's property" over land, air, inland surface waters, and underground resources such as minerals, oil, and water. It gave the president of Mexico the power to regulate water resource management.

Article 27 does not completely prohibit private management of water, since the president can grant concessions. While supporting the idea of public planning, however, it also was meant to constrain private ownership of water as a market commodity (Pablos, 1999, 95). Many have interpreted the Mexican Constitution of 1917 as formally granting a universal right to water and sanitation (Castro, 2004, 341). Eventually, amendments to Article 27 altered the status of water from a public good to an economic resource subject to the marketplace.

The Ministry of Water Resources initially handled Federal control of water, with agriculture playing a major role in policy formation. According to Robert R. Hearne, professor of Agricultural Economics, "Until the 1990s, these institutions maintained an orientation toward construction, irrigation development, and support for land reform. Because of this orientation, the Mexican government dedicated 80 percent of its agricultural budget toward the construction and maintenance of 81 Irrigation Districts. These systems were used to support the land reform *ejidos* [an area of communal land used for agriculture]" (Hearne, 2004, 3).

The institutional structure of developing and distributing water resources was centralist in post-revolutionary Mexico as were many other service functions and continued to develop as such through the mid- to late-twentieth century.

The propensity to centralize water was formalized under the Federal Law of Sanitary Engineering published in the Official Gazette of the Federation in January 1948. This law empowered the Secretariat of Hydraulic Resources (SRH) to manage drinking water and sewerage systems, while the federal government engaged in the construction of hydraulic works. That same year the General Directorate of Potable Water and Sewage (DGAPA) was established under the SRH to construct and operate most of the water supply and sewer systems in the cities. From these changes emerged other bodies to further control local water services further down the bureaucratic chain (Briseño and Sanchez, 2018).

This arrangement for public provision of water supply and sewerage in the hands of federal agencies (for the most part) lasted from the presidency of Miguel Aleman (1946–1952) at least through the presidency of Luis Echeverria (1970–1976). It was in line with government-owned and operated enterprises in a variety of economic sectors such as oil refining, food processing, sugar cane milling, and textile manufacturing. Between the late 1950s and 1960s, as the economy swiftly grew, the number of state-owned enterprises (SOEs) more than doubled. Supervision of these enterprises, however, became more complicated and the government responded with additional centralized control (Chong and Lopez-de-Silanes, 2004, 8).

Former Rhodes Scholar David Adler expressed criticism of the Mexican government which was shared by others:

> The History of Mexico's water is defined by overuse and under-preparation. Throughout the Mexican miracle, as the economic boom of the 1940s to the 1970s is known, industries continually overused aquifer reserves, as politicians sought to promote economic growth. As the boom wore on, Mexicans flooded the cities in pursuit of new economic opportunities...In Mexico's cities, where most of the growth was concentrated [between 1950 and 1980], water infrastructure could not keep up with increasing demand... Nowhere was this more visible than Mexico City..."
>
> (Adler, 2015)

Sociologist Jose Esteban Castro argued, "In practice, the enforcement of Article 27 became the object of protracted struggles throughout the twentieth century, and the universalization of urban water services ranked very low in the scale of priorities until the 1970s" (Castro, 2004, 329). Those programs that seemed most essential to economic growth were the focus of government promotion at the expense of services such as water provision, especially in the cities.

Beginning in the first half of the 1970s, dwindling private investment and stricter government restrictions consolidated the role of the public sector as the main source of capital investment in Mexico. As a result, the Federal government borrowed heavily and utilized income from oil to expand the number of SOEs under its control. In a haphazard manner, it took over companies that faced financial hardships or were of special interest to political leaders. By 1982, there were 1,155 SOEs in the country, putting heavy pressure on the economy (Chong and Lopez-de-Silanes, 2004, 9).

With growing economic stress, population growth, and demand for more services, successive federal governments since the 1970s proposed the decentralization of several programs to reduce costs, distribute jobs, and spread around patronage geographically. However, as Hearne argued, "Decentralization of Mexico's governmental institutions was initiated during a time of unchallenged presidential power and did not imply a deconcentration of influence...Decentralization was not intended to reduce presidential authority" (Hearne, 2004).

The National Water Law of 1972 established that private individuals could profit through a concession or license from government authority. It also set a priority order for access to water: domestic use, then agriculture, and finally industrial use. In 1971, the General Directorate of Operations of Potable Water and Sewage Systems (DGOSAPA) was created to supervise and operate the country's hydraulic systems. It included (in 1976) 34 delegations and a regional headquarters, 873 federal boards, 146 municipal committees, and 37 administrative committees. Unfortunately, this bloated bureaucracy did not have the authority to meet the needs of the exploding population, and its ineffectiveness was a factor in promoting decentralization. In 1976, the functions of the DGOSAPA were transferred to the Secretariat of Human Settlements and Public Works (SAHOP). The intention was to allow the SRH to focus on large hydraulic infrastructures, such as dams and irrigation districts (Briseño and Sanchez, 2018). Was this simply rearranging deck chairs on the *Titanic?*

In the late 1970s, privatization of water management (and other resources) became a major topic of interest worldwide, no less in Mexico. Development strategies turned to neoliberalism as an answer. One definition describes neoliberalism as "an economic policy model that puts great emphasis on property rights, rule of law, and free markets to maximize individual freedoms and lead society to its maximum potential." Geographer David Harvey more pointedly argued that in this model, "private enterprise and entrepreneurial initiative are seen as the keys to innovation and wealth creation" (Van Dusen, 2016, 6).

For Mexico, this would mean a substantial shift away from state management of resources to a program of decentralization and privatization.

For water, such policies were aimed at reducing the role of the federal government while increasing the involvement of the private sector. In fact, despite the steady increase in demand for water, state investment in water infrastructure had declined since the National Water Law (Van Dusen, 2016, 10).

In 1980, management of some potable water systems was transferred to the states to decentralize the service; some states did the same with municipalities. The trend toward water privatization, according to Marrero, "dates back to 1983 when then-President Miguel de la Madrid made changes in Article 115 of the country's Constitution which made water supply the responsibility of municipal governments" (Marrero, 2005). Responsibility for drinking water and sewage, in some cases, was relegated both to the states and to municipalities. Such moves not only transferred the infrastructure networks but also the management problems. Little mention was made about financing the systems. Few local governments were prepared for such a drastic change in policy or financial responsibility.

Since agriculture utilized about 80 percent of the country's water in the 1980s (Mexico was the seventh-largest irrigator in the world), decentralization had a major impact on nonurban water-supply systems as well. The federal government's substantially reduced capacity to manage the water system in the 1990s included the failure to improve water allocation, to answer questions of accessibility and efficiency, and to contain the actions of private sector players in water management (Van Dusen, 2016, 2).

Decentralization occurred gradually. In 1988 water in 21 of the 32 states was administered by state governments, 11 by municipal governments (Briseño and Sanchez, 2018). By 1996, 29 states had passed legislation authorizing private involvement through service contracts (Pablos, 1999, 116).

The water management system remained unwieldy going into the administration of Carlos Salinas de Gotari (1988–1994). The National Water Commission (CNA, within the Ministry of Agriculture and Water Resources, created in 1989) was the primary national body responsible for promoting decentralization (somewhat ironic) allowing for greater autonomy to state and local water agencies and the opportunity to transfer many (or all) functions to private sector operators (Barkin and Klooster, 2006, 12).

During the Salinas years, the size and scope of privatization in all areas reached a peak in Mexico. The government sold Mexico' telephone system, steel industries, roads and bridges, and about half of the commercial banks. Salinas also approved privatization of the social security system and state-owned commercial farms. Firms sold during the period represented more than 96 percent of all assets privatized (Chong and Lopez-de-Silanes, 2004, 10).

The 'Mexico model' of decentralization in the 1990s appeared to be a standard for other countries to measure against. The country's water policy, in line with other programs of the Salinas era, increasingly shifted away from the old strong-state control approach to decentralization and privatization in an attempt to create a new "water culture." One of the casualties of decentralization was effective environmental regulation.

In 1992, privatization in the water sector accelerated under the National Water Law. It was the most expansive reform in the nation's water policy, embedded in the National Development Plan. The new law was in line with global water policy, and dependent heavily on funding from the World Bank that assertively promoted privatization. In addition, new legislation by the mid-1990s separated land rights from water rights, making way for more private holdings. A number of cities—including Aguascalientes, Saltillo, and Mexico City—began to contract water management to transnational corporations such as Vivendi. All did not go well in some Mexican cities.

In the resort community of Cancun (where Mexico's first comprehensive private concession was let), Azurix—a subsidiary of the infamous Enron Corporation—became the first private company to run its water system. After its bankruptcy, the city turned to Ondeo, a subsidiary of Suez Corporation, which financed its purchase with a loan from the Mexican Public Works and Services National Bank. Investments promised under the arrangement did not materialize going into the twenty-first century. In 2000, the state government of Quintana Roo took over Cancun's water and sanitation services.

In Saltillo, the water system was contracted to a company owned in tandem by the municipality and Aguas de Barcelona. During the first two years, rates increased between 32 and 68 percent. In Aguascalientes, Vivendi assumed the contract and rates soared there as well (Marrero, 2005).

In Mexico City, there was a long struggle to import water from outside the city to meet the ever-growing demand. This high demand for water produced the moniker—"thirstiest city in the world." The city's water infrastructure had been falling apart for years with many broken water mains and thousands of breaks on primary and secondary systems. Water experts Cecilia Tortajada and Enrique Castelan argued that in trying to meet demand by successive governments exclusively through supply management and engineering solutions, "Environmental, economic and social policies associated with water management are mostly inadequate and insufficient, which is resulting in increasing deterioration in the environment, health and socioeconomic conditions of a population living in one of the largest urban agglomerations in the world." No long-term strategies have been developed (Tortajada and Castelan, 2003, 124).

The first substantial water reforms introduced in the Federal District (Mexico City) occurred in 1989, but the inability to constantly acquire adequate supplies of water and to repair the vast infrastructure led to calls for privatization in the earlier 1980s. With the call came many protests and much public unrest. In 1986, for example, people in the municipality of Naucalpan reported major interruptions in water supply and blamed profit-taking vendors. "People protested," Castro stated, "denouncing 'the voracity of water vendors' and the lack of protection against the unregulated market allocation of water supplies" (Castro, 2004, 336).

Figure 22.1 A farmer climbs a fence during a protest at Los Pinos Presidential house in
Mexico City on March 18, 2015. Farmers protested against the privatization of
water and demanded government policies that support the agricultural sector
and housing, according to local media. REUTERS/Alamy Stock Photo.

In 1993, the Federal District granted concessions to four private groups,
transferring a variety of responsibilities including metering and billing. But
even private sector approaches did not significantly reduce budget deficits by
the late 1990s, as costs and rates increased, and the public response to the
changes was essentially negative, including refusal to pay water bills (Castro,
2004, 339–340). Restless times over water management extended into
the first two decades of the twenty-first century. On March 22, 2015, for
instance, hundreds of protesters marched toward the National Water
Commission. One sign read *"El agua es nuestra, carajo!* Or "The water
is ours, goddamnit!" Another read *"El H20 no ese un negocio"*—"Water is not
a business."

With the public water utilities in disarray, private companies still attempted
to fill the gap in Mexico City and elsewhere, but with varying degrees of
success. Confusion reigned. Overall, between 1982 and 2003, SOEs dropped
from 1,155 to 210 (Chong and Lopez-de-Silanes, 2004, 9). In 2000, CNA had
decentralized many of its functions to 13 regional directorates dealing with
water. Privatization in the Federal District was unique in that private companies were part of a competitive service structure, while the government
retained control of the infrastructure. Debate continued as to whether private
sector concessions should be allowed in the Federal District.

Who manages the water system and how effectively in Mexico elicits no simple answers. The fundamental question of adequate supplies seems to be grounded in short-term decisions or no decisions at all. The use of bottled water remains essential in Mexico, estimated at 125 gallons per person annually. And private tankers—inadequate, expensive, and inefficient—deliver water to millions of Mexicans (Adler, 2015). Outside the cities, demands for massive quantities of water for irrigation exacerbate the country's water dilemma. As does the question of water quality and adequate environmental protection of the resource.

The question as to whether water privatization has succeeded or failed in Mexico depends on who you ask. What is clear is that while decentralization did not produce a uniform response to reform in water management, it shifted authority (and responsibility) away from the Federal government for better or worse. Urban water services in Mexico, according to Barkin and Klooster, "are delivered through a wide variety of administrative structures in different parts of the country." While the most common organizational form comes through a municipal department, most Mexicans get service through semi-autonomous (public sector) organizations (Barkin and Klooster, 2006, 18). By 2000 there were 360 operating organizations and 435 water utilities in Mexico (both mostly municipal).

The move to decentralization does not mean that water service problems have been addressed fully in Mexico, or that the debate over water as a right versus a commodity has been resolved. Briseño and Sanchez effectively argue that, "To speak of a crisis in the management of water for urban use in Mexico is to refer to a decentralization that did not bring the expected results...[The] water utilities in Mexico have not fulfilled the objectives for which they were created, which are mainly to provide water to the population at the lowest possible costs and limit waste by being financially self-sufficient" (Briseño and Sanchez, 2018).

Decentralization did not contemplate—or ignored—the limited or poor financial capacity of most municipalities in Mexico or their diversity. The cities also faced the same problems that the Federal government had faced and that some private sector actors also faced. Politics, along with bouts of corruption, always was intertwined into Mexico's water-use history (just like in every other country on earth). Some claimed, for instance, that when Vincente Fox Quesada was president, the Coca Cola Company received one of the largest amounts of water in Latin America. Coincidently, Fox had been the general manager of Coca Cola in Latin America. In 2014, Coca Cola's Disani water was sold for 46 cents, while the soft drink was sold for 26 cents.

Not all claims of partisanship or corruption fell on one side. Those supporting or deriding privatization sometimes made false claims about the success or failure of the private companies operating in Mexico.

By redefining the role of the state, decentralized governance of water resources was "a centerpiece of Mexico's neoliberal reform strategy" (Wilder and Lankao, 2006, 1977). And like elsewhere, as Barkin and Klooster argued,

"Water management provides a critical lens onto the development process" (Barkin and Klooster,2006, 1). The bottom line was that Mexico's mixed system of water governance largely failed the needs of its citizens and undermined the aspirations of the Revolution to make water a public good.

Notes and Further Reading

Adler, David (2015) "The War for Mexico's Water," *Foreign Policy*, July 31, online, https://foreignpolicy.com/2015/07/31/the-war-for-privatization-mexicos-water/.

Barkin, David and Daniel Klooster (2006) "Water Management Strategies in Urban Mexico: Limitations of the Privatization Debate," MPRA Paper No. 15423, *Munich Personal RePEc Archive*.

Briseño, Hugo and Antonio Sanchez (2018) "Decentralization, Consolidation, and Crisis of Urban Water Management in Mexico," *Water Science and Technology* 9 (July/August), online, http://www.scielo.org.mx/scielo.php?pid=S2007–24222018000400025&script=sci_arttext&tlng=en.

Castro, Jose Esteban (2004) "Urban Water and the Politics of Citizenship: The Case of the Mexico City Metropolitan Area during the 1980s and 1990s," *Environment and Planning A* 36, 327–346.

Castro, Jose Esteban (2006) *Water, Power and Citizenship: Social Struggle in the Basin of Mexico* (Oxford: Palgrave).

Chong, Alberto and Florencio Lopez-de-Silanes (2004) *Privatization in Mexico*, Working Paper, No. 513 (Washington, DC: Inter-American Development Bank).

Hearne, Robert R. (2004) "Evolving Water Management Institutions in Mexico," *Water Resources Research* 40 (December): 1–11.

Marrero, Carmelo Ruiz (2005) "Water Privatization in Latin America," Global Policy Forum, October 18, online, https://archive.globalpolicy.org/social-and-economic-policy/global-public-goods-1–101/46063-water-privatization-in-latin-america.html.

Pablos, Nicolas Pineda (1999) "Urban Water Policy in Mexico: Municipalization and Privatization of Water Services" (PhD Dissertation, University of Texas-Austin, December).

Scott, Christopher A. and Jeff M. Banister (2008) "The Dilemma of Water Management 'Regionalization' in Mexico under Centralized Resource Allocation," *Water Resources Development* 24 (March): 61–74.

Tortajada, Cecilia and Enrique Castelan (2003) "Water Management for a Megacity: Mexico City Metropolitan Area," *Ambio* 32 (March): 124–129.

Van Dusen, Richelle (2016) "The Politics of Water in Mexico City," University Honors Thesis, Portland State University.

Wilder, Margaret and Patricia Romero Lankao (2006) "Paradoxes of Decentralization: Water Reform and Social Implications in Mexico," *World Development* 34: 1977–1995.

Part VIII

Social Crises/
Environmental Injustices

Throughout the book, we have seen water episodes that examine, in some measure, environmental injustices—from the acequias in New Mexico to the treatment of indigenous peoples and others in the California Gold Rush, from "Levees-Only" in Louisiana to racism at swimming pools, and with water privatization in Mexico. In this section, we explore recent iconic cases of environmental injustice in the Flint, Michigan, water crisis and the maquiladoras on the Mexico–U.S. border. In addition, a third episode looks at the debate over hydraulic fracking in Mexico which raises questions about our energy future tied to fossil fuels.

The Flint water crisis flashed across the media and renewed debate over the safety of our water-supply systems. But it was the people who suffered in the crisis and the apparent indifference of the government that was at the heart of the story. Flint became big news not simply because of the unique and disturbing character of the incident which began in 2014, but because similar events arose before and could (or would) happen again.

In 2003, Virginia civil engineer Marc Edwards—the same Marc Edwards who researched the Flint case—conducted research into leaks in copper pipes in the DC area under funding from the District of Columbia Water and Sewer Authority (WASA). He found high lead concentrations in the pipes and realized that WASA had been giving out misinformation about it. WASA refused to issue a new memo on the problem and threatened to terminate Edwards' funding. In 2004, the U.S. Environmental Protection Agency ruled that WASA was in violation of federal regulations, and in 2009 a congressional hearing revealed that the Center for Disease Control and Prevention had earlier produced a flawed report on the WASA lead case (Burke, 2016).

Other contamination experiences occurred in places such as Milwaukee, Toledo, and elsewhere. For First Nations people in Canada—going back to early colonialism—water advisories were chronic. In 2015, there were drinking-water advisories in 126 First Nations. Many indigenous communities in Canada, according to one study, "live with high-risk drinking water systems and drinking water advisories and experience health status and water quality below that of the general population" (Bradford et al., 2016).

DOI: 10.4324/9781003041627-31

The Flint case had all the markings of a classic incident of environmental injustice. If there was a small ray of hope it was the response of Flint citizens to their misfortune. Environmental historians Sara B. Pritchard and Carl A. Zimring contended, "In such environments, residents learn the risks of the landscapes and adapt accordingly. The adaptations are themselves not without hazards, but the knowledge humans develop in responding to toxic environments resonates with the adaptations residents in Flint, Michigan, made in boiling water, using bottled water, and otherwise responding to the risks of bacterial contagion and lead poisoning..." (Pritchard and Zimring, 2020, 127–128) This does not mean that the Flint citizens resigned themselves to their plight, because they did not as numerous protests demonstrated. But in showing resilience, they confronted what was otherwise a dire future.

The maquiladoras along the Mexico–U.S. border unmask a case of ongoing, relentless pollution sustained by disregard for working conditions in the myriad factories and adjacent communities among thousands of Mexican laborers. Since the majority of those workers are women, this story takes into account environmental injustice disproportionately experienced along gender lines. The intent of the maquiladoras program starting in the 1960s was to pump foreign investment into Mexico to buoy up its sagging economy. Profits not people's welfare has dominated this enterprise even after the passage of the North American Free Trade Agreement initiated in the 1990s. The Mexican government has been saddled with the blame for conditions in the maquiladoras because of weak environmental laws and lax inspections. But a sizable degree of responsibility falls on the non-resident factory owners themselves.

In many respects, this case is unique because the environmental problems that the maquiladoras have created fall on the shoulders of those who promoted them and those who operate them across a common border. Of course, both the United States and Mexico seek vast economic gain, but accept different levels of responsibility for the condition and output of the factories. Caught in the middle are the workers and others living in a maquiladora communities. The cooperating governments have failed to work out practical solutions to bad working conditions and layers of environmental pollution.

Other border issues dealing with questions of water rights, water access, and pollution often begin at different starting points than the story of the maquiladoras. The essay on the Fraser River, for instance, demonstrated how an unusual set of circumstances resulted in an accommodation over salmon and hydropower between the United States and Canada.

The tri-state water wars among Alabama, Georgia, and Florida—although not represented in this book—is a confusing story in its own right. For more than 30 years, the three states have disputed the use of shared river basins—the Apalachicola-Chatahhoche-Flint and the Alabama-Coosa-Tallapoosa. The complicating factor is that the conflict has been fought over multiple needs—drinking water, power generation, agriculture, navigation, and recreation—where hydrology, topography, geography, and politics mix to make resolution nearly impossible (Atlanta Regional Commission, 2021).

The last water episode moves away from environmental injustice toward energy exploitation and the possible environmental risks. We have explored these issues earlier, especially in the cases dealing with waterwheels and steam engines, Niagara Falls, the Houston Ship Channel, the Fraser River, and Ixtoc 1. Fossil fuel extraction is at the heart of this case, as in the story of the Ixtoc 1, and the outlines bear close resemblance. In both instances, oil companies recognized new opportunities to acquire what seemed to be inaccessible supplies of oil and gas by employing innovative technologies. Controlling those resources became an end in itself no matter what the prevailing market for oil and gas might be. Also, in both cases, the potential environmental risks of such ventures were downplayed or ignored—literally the cost of doing business. And, in different ways, water plays a crucial role in the success (or failure) of the enterprises.

What makes the fracking debate in Mexico unique is that despite the country's need for an expansion in its economy, a variety of factors have stymied large-scale commitment to hydraulic fracking. The location of the biggest and best area of exploitation—the Burgos Basin—suffers from an array of challenges, which include the aridity and lack of water supplies in that part of Mexico, the need for foreign investment dollars, cartel activity along the border, and the limited success of PEMEX in stimulating more oil and gas development. The wild card in recent years has been the presidency of Andres Manuel Lopez Obrador, a populist, with deep skepticism of fracking. Ironically, Lopez Obrador wanted to expand traditional oil and gas exploration, which brings into question the degree to which his objections to fracking are built on environmental concerns. To frack or not to frack in Mexico is linked to several local and historical factors unique to that country, while the environmental issues, concerns over water use, and financial costs are more universal.

Notes

Atlanta Regional Commission (2021) "Tri-State Water Wars Overview," September 23, online, https://atlantaregional.org/natural-resources/water/tri-state-water-wars-overview/.

Bradford, Lori E.A., et al. (2016) "Drinking Water Quality in Indigenous Communities in Canada and Health Outcomes: A Scoping Review," *International Journal of Circumpolar Health* (July 29), online, https://pubmed.ncbi.nlm.nih.gov/27478143/.

Burke, Katie L. (2016) "Flint Water Crisis Yields Hard Lessons in Science and Ethics," *American Scientist* 104, May–June, online, https://www.americanscientist.org/article/flint-water-crisis-yields-hard-lessons-in-science-and-ethics.

23 The Flint Water Crisis

This headline ran on the front page of the *Houston Chronicle* on October 12, 2021: **"Experts fear Fifth Ward is one of 'hundreds of Flints.'"** The article detailed how environmental advocates and university researchers found lead in water, paint, and soil in Houston's Fifth Ward, a mostly Black and Latinx neighborhood on the city's northeast side. "[Lead] has emerged as a major environmental justice issue," it continued, "gaining renewed attention after the Flint, Mich., water crisis began in 2014, when lead seeped into the city's water after it switched its supply and pipes began to corrode."

The water crisis in Flint, an industrial suburb of Detroit, Michigan, occurred seven years before the *Chronicle* story and clearly achieved iconic status, coming to represent widespread environmental injustice yet to be rectified. The story of Flint and its water controversy is quite complex, however, attributed to racial injustice, deep forces of history, and governmental mismanagement. To what degree did each of these causes contribute to the events in Flint and their outcome is the focus of this essay.

The shorthand version of the crisis begins with the observation that Flint had a long history of racial segregation, growing poorer and falling into physical decay because of deindustrialization and white flight. On April 16, 2014, a state-appointed emergency manager and his staff, seeking to cut costs in the financially weakened city, decided to temporarily switch the water supply to the Flint River while an alternative system was being completed. The water in the Flint River was highly toxic, especially since no corrosion controls were implemented before entering the city's homes and businesses.

Citizens immediately began to complain about the taste, appearance, and smell of the water, while the government assured them that the water was reliable. Thus, began a long process to determine the safety of the water, to address potential shortcomings of the system, and ultimately to assess responsibility—and blame—for the incident itself. Along the way, the specific problem of water impurity blossomed into a larger debate over government mismanagement, citizens' rights, and racial inequality.

The City of Flint's history before 2014 is important context for the crisis, some even arguing that imbedded in it was the cause of the crisis itself more than simple mismanagement. At one point, Flint was a manufacturing hub of

DOI: 10.4324/9781003041627-32

General Motors Corporation (GM). But as scholars Richard Casey Sadler and Andrew R. Highsmith stated, "Over the past three-quarters of a century [before 2016], a harsh mix of divestment, 'white flight,' metropolitan political fragmentation, and persistent racial discrimination transformed this once economically vibrant although deeply divided city into one of the poorest, most racially segregated metropolitan regions in the United States." In this setting, revenue steadily declined and the infrastructure decayed, which ultimately set the stage for a state takeover of the municipal government (Sadler and Highsmith, 2016, 144).

It was an old story. As deindustrialization decimated Flint's economy, thousands of white residents fled the city for new suburbs, where they created independent governments and broke relationships with the inner city where many African Americans and the poor remained. The depopulation weakened the tax base, and insulation of many whites in the suburbs meant, among other things, that basic public services in Flint were neglected. The fragmentation of the metropolitan area further perpetuated segregation and racial/economic inequality (Sadler and Highsmith, 2016, 144).

Before World War II, although the automobile industry was thriving in Flint and population grew, racial segregation in housing was clearly evident. Many black families lived in crowded, polluted, and rundown areas near the GM factories. Inequity in the workplace along with racial discrimination accounted for that reality. Between 1919 and 1933, GM helped perpetuate segregation by constructing 3,000 new whites-only housing units on the west side of Flint. Indeed, most new housing starts in the period took place in segregated neighborhoods.

But even the city's periphery in the 1940s and 1950s was not attractive enough for whites who could afford to leave the central city. To retain GM's presence, municipal leaders began investing in services in the new suburbs for those whites migrating there and for various industrial land uses. The people remaining in the central city saw little if any new investment. GM proposed a solution to the uneven development of the metropolis called "New Flint," to coordinate growth and the financing of new services. Many suburban entities balked, especially because they believed they already were overburdened with debt and opposed regional governance. This stance, Sadler and Highsmith concluded, steered a course "toward deindustrialization and sociopolitical fragmentation that ultimately led to Flint's 2011 financial emergency and the subsequent water crisis" (Sadler and Highsmith, 2016, 147).

While this assessment may be insufficiently complex to explain conditions leading directly to the water crisis, it is suggestive of Flint's downward trajectory—one that was not unlike what was happening throughout the so-called Rust Belt—which widened the gap between black and white, rich and poor (Pauli, 2019, 61–67). Flint was now a majority-black city; overall having 62.6 citizens of color. Almost 42 percent of its people lived below the poverty line. The median income in Flint in 2014 was less than half of the median income for the United States as a whole (Butler et al., 2016, 94).

In response to the water crisis in 2014, the Michigan Civil Rights Commission issued a report in 2017, reinforcing what it believed to be systemic racism in events leading to and following the crisis. "Historically, Flint's community of color was long relegated to substandard housing, education, and job opportunities. Initially, the segregation was the result of both private discrimination...and government policies..." (Michigan Civil Rights Commission, 2017, 3).

Figure 23.1 Cast away water pipes in the Flint area. Getty/AYEHAB.

Along with the momentum of the past, the historic state of the Flint River is an important part of the crisis story. From 1974 to 2014, water from Lake Huron had provided the city's water supply. The Flint River was not a viable source of potable water for several reasons. Water-treatment specialists had not been able to treat Flint water successfully, especially because of the variety of wastes that polluted it. These included natural biological waste, treated human and industrial waste, untreated water, illegal substances dumped in the river, and contaminants from runoff. The Flint River also was warmer than Lake Huron (raising levels of bacteria) and flowed inconsistently, especially in summer. Even with the departure of heavy industry, the river remained very polluted. As one writer noted, "It is hard to imagine why anyone familiar with the river's history would ever decide to use it even as a temporary water source. But they did" (Carmody, 2016).

The history of Flint's declining economic fortunes and the state of the Flint River are part of the Flint water crisis, but they do not offer a complete explanation. While they provide context and momentum, decisions made in 2011 and thereafter—growing out of a variety of circumstances—came to a head in 2014 when the decision was made to temporarily utilize the Flint River for the City of Flint's drinking water. That decision exposed government indifference and mismanagement and intensified public mistrust in city and state leadership.

Flint's mounting financial woes at the turn of the twenty-first century were exacerbated by the decision of the State of Michigan in 2003 to sharply reduce the program of revenue sharing with municipalities, leading to a decline in funds going to the cities in the range of $900 to $250 million annually. Flint stood to lose $54.9 million between 2003 and 2014. In addition, in 2011 the legislature passed an emergency manager law that allowed state officials to take over municipal finances and directed the emergency managers to cut spending where possible. (Votes overturned this law the next year, but the legislature passed an even more stringent law with no provisions for a referendum.) Flint became the first city in Michigan to face state receivership under the new law. Many considered these actions punitive, and clearly an expression of a loss of faith in the ability of municipalities to correct or solve their own financial problems (Sadler and Highsmith, 2016, 149). Michigan Republican governor Rick Snyder appointed the first in a series of four (unelected) emergency managers on November 29, 2011, replacing the (elected) city government in Flint.

Announced as a cost-saving measure for the city, officials on the local, county, and state levels (led by the state-appointed emergency manager Ed Kurtz) switched Flint's water supply from Lake Huron to the Flint River. The Flint Water Service Center (FWSC) substituted distributing Lake Huron water provided by the Detroit Water and Sewage Department (DWSD) to utilizing Flint River water to be treated through its own facility. There had been perpetual disagreements between DWSD and the City of Flint over rates; Flint residents were paying 2.5 times the national average. Some argued, most obviously, that the water utility simply was ripping off Flint customers (Paul, 2019, 1).

Kurtz and his supporters decided to work with the Karegnondi Water Authority (KWA) which agreed to build a new pipeline to Lake Huron which, they argued, would provide less expensive water than DWSD and save

$200 million over twenty-five years. Social Science professor Benjamin J. Pauli suggested that a switch in water supply (which would be very lucrative to KWA) also was attractive because it would allow the City of Flint to better utilize its own water treatment plant and gain more control over the governance and operation of its water supply system.

Even at this early juncture, however, community activists were convinced that KWA was a scam operation, and not an improvement over DWSD (Pauli, 2019, 82–88). Representatives of the Michigan Department of Environmental Quality (MDEQ) insisted that citizens "shouldn't notice any difference" in the Flint River water other than an increase in hardness which might reduce lather when washing up. They later would make misleading claims about the lead problem. City officials also continued to respond optimistically about the water quality (Pauli, 2019, 3, 5).

As a temporary fix, since the KWA pipeline was not completed (and DWSD immediately terminated the contract with Flint), it was decided to draw water from the Flint River. This was the first time since the 1960s that the river was used for such a purpose (Robertson, 2020). Mayor Dayne Walling declared "It's regular, good, pure drinking water, and it's right in our backyard" (Kennedy, 2016). The problem was that officials did not stipulate—and apparently hid the fact—that the water should be subject to corrosion controls (which it was not) in order to be utilized properly. The "temporary" use of Flint River water lasted 18 months.

As the crisis began to mount state officials claimed that they were only following the lead of locals in using the Flint River, that is, the people of Flint were only courting water problems themselves. Yet, the emergency management system that controlled local affairs at the time offered a clear indication that demand for changing to Flint's water supply did not have local government nor grassroots origins, but in state-generated financial concerns (Pauli, 2019, 91–95).

Change came quickly. In May 2014, *Escherichia coli* and total coliform bacteria were detected in Flint's water. By August 2015, the water exceeded the limits for *E. coli*, leading to three boil-water alerts in a period of 22 days and an increase in chlorine being added to the water. The level of lead in Flint's water surged after the change was made, despite claims from officials that the new supply was not a cause for concern.

Almost immediately complaints swamped the city about the discoloration of the water, its odor, and its metallic taste; bottled water sales swelled (even the state began buying bottled water for its employees!). People talked about having skin rashes, hair loss, and an array of mysterious symptoms.

On October 13, General Motors announced that it would no longer use river water at its engine plant, fearing it would cause serious corrosion problems, and then turned to Lake Huron for its water. In January 2015, Flint was found in violation of the Safe Drinking Water Act because of high levels of total trihalomethanes (TTHM) in the water—carcinogens composed of byproducts of chlorine and organic matter (Clark, 2016, 100).

Ten months after the April 2014 decision, water samples collected from the Flint River showed progressively rising levels of lead in the water as well as more discoloration. The cause: No corrosion control added to the drinking water source. As a result of those samples, engineers at Virginia Tech stated,

> Forensic evaluation of exhumed service line pipes compared the water contamination "fingerprint" analysis of trace elements, revealed that the immediate cause of the high water lead levels was the destabilization of lead-bearing corrosion rust layers that accumulated over decades on a galvanized iron pipe downstream of a lead pipe.

Analysis of blood lead data in Flint children in September 2015 showed serious spiking lead numbers. A declared state of emergency "likely averted an even worse exposure events" (Pieper, Tang, and Edwards, 2017). A team of researchers led by Dr. Mona Hanna-Attisha at Flint's Hurley Hospital had shown through a study that lead from the Flint River was present in the city's children, putting them at risk for a variety of developmental problems (Pauli, 2019, xi–xii).

Dr. Marc Edwards, a member of the team at Virginia Tech who tested water in hundreds of homes, asserted, "The levels that we have seen in Flint are some of the worst that I have seen in more than 25 years working in the field" (Kennedy, 2016). Although some officials tried to discredit the work done at Virginia Tech, media reports on Flint's toxic water produced national headlines. Medical experts believed that lead exposures were likely to have detrimental health impacts for many years.

On December 14, 2015, Mayor Karen Weaver declared a local state of emergency, followed by Governor Snyder's declaration of a state of emergency on January 5. Both requested immediate federal assistance. On January 16, President Barack Obama declared a federal state of emergency authorizing the Federal Emergency Management Agency (FEMA) to coordinate relief efforts. During this frenzied time, investigations into responsibility for the crisis began on the state and national levels. To make matters worse, on January 16 the governor and officials from the Michigan Department of Health and Human Services announced that there had been a major outbreak of Legionnaires' disease (a severe lung infection caused by the bacteria *Legionella*) going back to June 2014. Eighty-seven cases were reported with twelve deaths possibly linked to the switch to Flint River water.

Protesters and activists in Flint were outraged by the health risks they were facing which intensified their growing mistrust of government. Pauli argued persuasively that citizens were fighting for "water justice" encompassing "everything from securing safe, affordable water, to replacing damaged infrastructure, to declaring water a human right, to community revitalization. The broader idea of 'environmental justice,' too was very much in the air" (Pauli, 2019, 14). He added that the activists' conviction was that "the injustice of the water crisis was the product of a prior crisis of 'democracy.'"

What the activists felt was lost "when emergency management was imposed by the state," he added, "was the right of residents' elected representatives at the local level to have any say over decisions about Flint's water (among other things)." The right of representative democracy was not available to them, or, at best, ineffective (Pauli, 2019, 14–15).

As the crisis dragged on, grassroots protest grew (with groups like the Coalition for Clean Water), the call for accountability intensified, and demands for long-term solutions to the water crisis itself were repeated time and again. Water activist Melissa Mays stated, "We just want to live normally, and actually be able to drink the water that comes out of our tap safely, with no concerns, like normal people" (Robertson, 2020).

Residents filed a class-action lawsuit in which nine officials—including Governor Snyder—were charged with 34 felony counts and seven misdemeanors. More charges followed. Thousands of Flint citizens filed individual lawsuits against the state. There have been more than 21,000 claims of damage or injuries. However, five years after the crisis began no one had been sentenced to prison. Several officials in office in 2014 are now gone.

In November 2021, a federal judge approved a $626 settlement (not including legal fees) in civil claims, which the State of Michigan will pay victims of the crisis, largely to be designated for children of Flint. It is one of the largest in state history. Yet the children's long-range health is not clear (*New York Times*, July 29, 2021; NPR, November 11, 2021).

While state and local officials were culpable in the Flint water crisis, the federal government had not bathed itself in glory. Among other things, there were weaknesses in the implementation and oversight of the U.S. Environmental Protection Agency (EPA), especially with respect to its lead and copper regulations (Katner et al., 2016, 109). At one point in the crisis, an EPA official purportedly stated, "I don't know if Flint is the kind of community we want to go out on a limb for" (Clark, 2018, 147).

The water source for Flint switched back from the Flint River to Lake Huron (treated in Detroit). In 2019, Flint water met federal standards (for levels of lead and copper). But suspicions continued to run deep and many people still used bottled water for cooking, drinking, and bathing despite the fact that the state's bottled water distribution centers closed down in 2018. Work to repair all pipes is not yet completed, stalled most recently because of the coronavirus pandemic. In 2019, more than 8,000 service lines were replaced, but several thousands more needed to be examined (*New York Times*, April 25, 2019).

Was the Flint Crisis an environmental justice event? Most certainly, but it had its own unique characteristics. Beginning in the 1980s, environmental justice advocates reinvigorated the demand for civil rights in a new political arena. They focused on the disproportionate exposure to health and pollution risks faced by people of color at home and in the workplace. Such incendiary terms as "environmental racism" and more clinical academic definitions such as "environmental equity," were set aside and replaced by the concept of "environmental justice." The latter term offered a more positive, more inclusive concept to interject into the political discourse and suggested a broader objective—justice—than simply the eradication of racist actions.

Sociologist Paul Mohai, a respected leader in the Environmental Justice Movement, stated, "Indeed, the Flint Water Crisis is an extraordinary example of environmental injustice affecting an entire city of 100,00 residents..." He agreed with the conclusion of Governor Snyder's appointed Flint Water Advisory Task Force (FWATF) that the water crisis

> ...lead us to the inescapable conclusion that this is a case of environmental injustice. Flint residents, who are majority Black or African American and among the most impoverished of any metropolitan area in the United States, did not enjoy the same degree of protection from environmental and health hazards as that provided to other communities.
>
> (Mohai, 2018, 31)

Mohai stated that injustice occurred on several levels: Distributive injustice, where a concentration of poor people and people of color suffer in a space that also suffers from contamination. Procedural injustice, in which residents are not given a meaningful voice in the decisions that impact their community. And social injustice, where problems are imbedded in a large social context beyond the event itself. All of these, he asserted, required corrective action to repair the harms caused by the Flint water case. Mohai testified before the task force. "I responded that I doubted we would find anyone confessing that there was an *intent* to do harm in Flint because of the presence of a large number of poor people and African American residents. What stands out in the Flint water crisis, however, is the apparent *indifference* and *lack of concern* that harm might be created" (Mohai, 2018, 34).

The Michigan Civil Rights Commission was most emphatic: "The people of Flint have been subjected to unprecedented harm and hardship, much of it caused by structural and systemic discrimination and racism that have corroded your city, your institutions, and your water pipes, for generations" (Michigan Civil Rights Commission, 2017, v). It remarked at several points in the report that "implicit bias" was clearly evident in the Flint case and throughout the country at large. But, it added, the racialized outcome experienced in Flint also had white victims. For the white families who remained in the city and did not flee to the suburbs, they were "victims of racism by association" (Michigan Civil Rights Commission, 2017, 85).

Journalist Anna Clark concluded, "One decision after another, one policy after another: [officials] were colorblind. Nobody explicitly argued that Flint should settle for water with high lead levels because it is mostly an African American city or because so many residents are poor. And yet" (Clark, 2018, 205).

One scholar argued that what happened in Flint was not due to racial disparity because water had been consistently devalued by the government in Michigan. Rural people had suffered as much as urban residents. "The poor, whether rural or urban, suffer disproportionately when their valuation of resources is not considered" (Clark, 2016, 99). This argument seems to suggest that environmental injustice, without the racial overtones, is injustice just the same.

Environmental justice discourse traditionally has been linked to race and to a somewhat lesser extent to class since its inception. But as more incidences become public the definition of environmental justice has and will continue to broaden. Race and class were at the core of Flint's water crisis, but the story of Flint while unique as to details is but an example of a much more universal problem. Pauli, seeking a hopeful—or maybe plaintive—note, stated,

> Flint became the centerpiece of a reinvigorated national discussion about aging infrastructure and served as a wake-up call about the lingering presence of lead in the urban environment. According to an oft-used metaphor, Flint was the canary in the national coal mine, a foreboding of crises to come if water systems were not upgraded and regulatory standards tightened. Many Flint residents took pride in the role the city played in raising this kind of awareness and potentially preventing future crisis.
>
> (Pauli, 2019, 33)

Stories about lead contamination gained traction in 2020–2021 for a variety of reasons including the announcement that the Joseph Biden administration proposed to earmark $45 billion in an omnibus infrastructure bill to eliminate lead from water because there were "hundreds of Flints all across America." Texas Southern University professor and a stalwart of the Environmental Justice Movement, Robert Bullard, asserted that the Fifth Ward [in Houston]—like in the Flint case—is a "classic example" of a community in need of investment for a variety of environmental and health problems. "You can see," he added, "the communities where the planning or the lack of planning has made communities much more vulnerable to heat, pollution, flooding—you start naming the environmental health disparities and you can map that and show that there is an economic dimension, there's a social dimension and there's a racial dimension."

To understand the importance of the Flint water crisis, however, is not only to view it as a harbinger of the future but also as a product of the past. Historian Chris Sellers aptly noted, "Long before that fateful decision … to turn the Flint River for the city's drinking water, pipes made of lead had threaded throughout the city's underbelly. Flint shares this historical legacy with thousands of other cities, suburbs, and towns across our country, and most likely this is not the first time, even in Flint, that these pipes have conveyed tiny amounts of the toxin into homes and children" (Sellers, 2016).

Notes and Further Reading

Butler, Lindsey J., Madeleine K. Scannell, and Eugene B. Benson (2016) "The Flint, Michigan, Water Crisis: A Case Study in Regulatory Failure and Environmental Injustice," *Environmental Justice* 9, 93–97.

Carmody, Tim (2016) "How the Flint River Got So Toxic," *The Verge*, February 26, online, https://www.theverge.com/2016/2/26/11117022/flint-michigan-water-crisis-lead-pollution-history.

Clark, Anna (2018) *The Poisoned City: Flint's Water and the American Urban Tragedy* (New York: Metropolitan Books).

Clark, Karen (2016) "The Value of Water: The Flint Water Crisis as a Devaluation of Natural Resources, not a Matter of Racial Justice," *Environmental Justice* 9, 99–102.

Flint Water Advisory Task Force (2016) *Final Report*, Office of Governor Rick Snyder, State of Michigan, online, https://www.michigan.gov/documents/snyder/FWATF_FINAL_REPORT_21March2016_517805_7.pdf.

Highsmith, A.R. (2015) *Demolition Means Progress: Flint, Michigan, and the Fate of the American Metropolis* (Chicago, IL: University of Chicago Press).

Katner, Adrienne, et al. (2016) "Weaknesses in Federal Drinking Water Regulation and Public Health Policies that Impeded Lead Poisoning Prevention and Environmental Justice," *Environmental Justice* 9, 109–117.

Kennedy, Merrit (2016) "Lead-Laced Water In Flint: A Step-By-Step Look At The Makings Of A Crisis," *NPR*, April 20, online, https://www.npr.org/sections/thetwo-way/2016/04/20/465545378/lead-laced-water-in-flint-a-step-by-step-look-at-the-makings-of-a-crisis.

Michigan Civil Rights Commission (2017) *The Flint Water Crisis: Systemic Racism Through the Lens of Flint*, February 17, online, https://www.michigan.gov/-/media/Project/Websites/mdcr/mcrc/reports/2017/flint-crisis-report-edited.pdf?rev=db527d0e6c4042 54892c84c907988934.

Mohai, Paul (2018) "Environmental Justice and the Flint Water Crisis," *Michigan Sociological Review* 32 (Fall): 1–41.

Pauli, Benjamin J. (2019) *Flint Fights Back: Environmental Justice and Democracy in the Flint Water Crisis* (Cambridge, MA: MIT Press).

Pieper, Kelsey J., Min Tang, and Marc A. Edwards (2017) "Flint Water Crisis Caused By Interrupted Corrosion Control: Investigating 'Ground Zero' Home," *Environmental Science & Technology 51*, January 10, 14.

Pritchard, Sara B. and Carl A. Zimring (2020) *Technology and the Environment in History* (Baltimore: Johns Hopkins University).

Robertson, Derek (2020) "Flint Has Clean Water Now, Why Won't People Drink It?" *Politico*, December 23, online, https://www.politico.com/news/magazine/2020/12/23/flint-water-crisis-2020-post-coronavirus-america-445459.

Sadler, Richard Casey and Andrew R. Highsmith (2016) "Rethinking Tiebout: The Contribution of Political Fragmentation and Racial/Economic Segregation to the Flint Water Crisis," *Environmental Justice* 9, 143–151.

Sellers, Chris (2016) "Piping as Poison: The Flint Water Crisis and America's Toxic Infrastructure," *The Conversation*, online, https://theconversation.com/piping-as-poison-the-flint-water-crisis-and-americas-toxic-infrastructure-53473.

Sellers, Christopher, et al. (2019) "The Flint Water Crisis: A Special Edition Environmental and Health Roundtable," *Edge Effects* (October 12), online, https://edgeeffects.net/flint-water-crisis/.

Warren, Christian (2001) *Brush with Death: A Social History of Lead Poisoning* (Baltimore: Johns Hopkins University Press).

24 Maquiladoras and Water Pollution

The Mexico–United States border of almost 2,000 miles has been the site of many water-related issues between the two countries. The Rio Grande River, which serves as part of that border, has been for many years a source of contention over water rights, but also a frontier for economic opportunity and resource exploitation. In northeastern Coahuila state in Mexico, the Burgos Basin sits across the border to the north and south of Brownsville, Texas. Hope has run high in recent years through fits and starts that Mexico could tap the vast shale oil and gas deposits there through a systematic program of hydraulic fracturing, which demands significant amounts of fresh water.

Another controversial issue—and the subject of this essay—are the maquiladoras (foreign-owned industries) which also promised renewed economic growth for Mexico, but generated extensive pollution, especially linked to water and wastewater. The maquiladoras originated along the border in the 1960s, exploding in number especially in northern Mexico during the 1970s.

The term "maquiladora" comes from the Spanish word *maquila*—the process of grinding wheat into flour in medieval Spain. It also can mean the toll paid on grain or flour to the miller or the lord of the manor for grinding grain. The modern meaning suggests a manufacturing operation processing raw materials and component parts into finished products to be sold in countries other than where they were manufactured (the market for goods has changed somewhat over time to allow for some products to remain in Mexico). Products well-suited to the maquiladoras include electronics, electrical equipment, ceramics, toys, and automobile parts.

Between 1993 and 1998, the maquiladoras accounted for 41.5 percent of average Mexican export value, and between 1994 and 2000 their share of foreign direct investment grew from 6 percent to 21.4 percent. The maquiladoras also generated pollution in many forms on both sides of the border, especially water pollution and the production of hazardous wastes. The need for fresh water by some of the industries is no less great than for fracking companies, and the resulting water pollution is no less serious. Environmental justice concerns in the factories and in the border towns intensified as the maquiladoras multiplied. Women workers in particular have been a key target of such problems.

DOI: 10.4324/9781003041627-33

For the Mexican government, the maquiladoras represented an expedited way to attract foreign investment by establishing a free-trade zone along its border with the United States. The majority of the businesses operating under this program are U.S.-owned transnational corporations. In 1970, Mexico had 72 such factories; by 1979, there were 620. In 2015, the number soared to about 3,000 maquiladoras along the northern border.

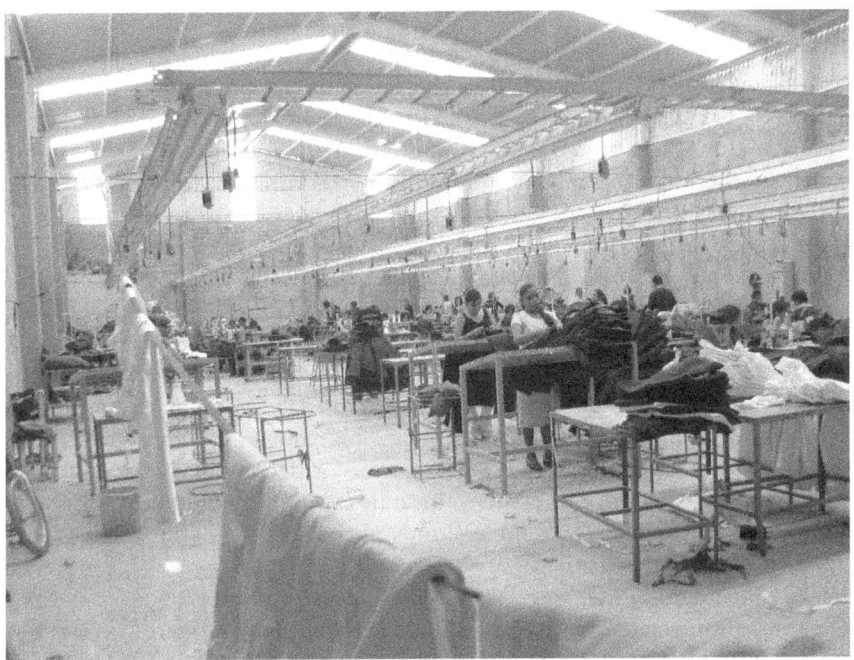

Figure 24.1 A Maquiladora in Mexico (May 1, 2007). Public domain. Wikimedia.

The roots of the maquiladoras go back to the Bracero Program, initiated by the U.S. government in 1942. Under the program, Mexican workers could migrate to the United States to help alleviate the domestic labor shortages caused by World War II. The program was terminated (unilaterally) in 1964, and thousands of Mexican workers were returned to Mexican border towns, which created its own set of challenges. Thousands of Mexicans had or were leaving the interior farmlands as well to find jobs in the cities. The over-crowding and rising unemployment at the border led the Mexican govern-ment to institute the National Border Industrialization Program (1965) to promote economic development and employment opportunities In doing this, it established a tariff-free zone along the border to promote investment (Grineski et al., 2015; Kotvis, 1996, 51; McDowra, 2020).

The establishment of the maquiladoras was attractive to outside companies because they promised cheap labor costs and lax labor and environmental regulations. Even as environmental regulations became more stringent in Mexico, there was little vigorous enforcement. The 1983 La Paz Agreement between the United States and Mexico was a pact to protect, conserve, and preserve the environment along the border. It sounded good at the time but resulted in few major results. For its part, the U.S. Environmental Protection Agency (EPA) had limited jurisdiction on the Mexican side of the border, even in cooperation with Mexico's *Secretaria de Desarrollo Urbano y Ecologia* (SEDUE).

Part of the reason for tepid results from enforcement was the expectation of considerable economic gain from the numerous plants operating along the border—financial success that Mexico desperately needed and American companies wanted. Sometimes corruption and bribery allowed officials to turn their heads away from obvious violations. It was common for U.S. authorities to blame health and environmental problems on lax Mexican laws, economic crises there, or the unwillingness of the Mexican government to invest in enforcement measures. The truth is, the American corporations operating across the border were massive contributors to those problems, several avoiding paying disposal costs or simply dumping wastes.

The environmental stresses along the border did not begin with the maquiladoras. Population growth spurred by industrialization put further pressure on water resources, waste management, and human health systems. Raw sewage had been dumped into watercourses for decades and sanitation infrastructure was faulty or nonexistent. In some locations, pollution was rife from earlier mining activities and mineral smelting. As of 1997, 40 percent of those people on the Mexican side of the border lacked sewers, potable water, or both. On the U.S. side, several hundred thousand people living primarily in colonias (unregulated housing developments or slums) also were without basic services (Clifford and Sheridan, 1997).

As the maquiladora system grew, the need for potable water and increasing pollution of watercourses grew hand in hand. Water quality has been a growing problem for nearly all Mexican border communities. Many people drew water from surface sources that had been polluted upstream or from aquifers contaminated with human sewage or chemical wastes. Water scarcity is a chronic problem, especially when the maquiladoras—such as textile and garment industries—placed heavy demands on available water sources. Some companies purchased private water rights as they did in the fracking businesses operating in Mexico. The maquiladoras also threatened the water supply by using city drainage systems or failing to control the effluents they produced (Williams and Homedes, 2001, 328).

As reported in the *Los Angeles Times* in 1997: "As border cities grow, ground water in some areas is being depleted at alarming rates. The main aquifer serving El Paso and the neighboring Ciudad Juarez is dropping at a rate of eight to 10 feet a year, Texas officials say." Juarez was still without a wastewater treatment plant at the time, and about 400,000 people in colonias (mostly in Texas) did not have drinkable water or sewer systems (Clifford and Sheridan, 1997).

The most vexing environmental problem facing the maquiladoras was the creation and disposal of hazardous wastes. Such wastes made their way into nearby groundwater, surface water, ground, and air, and were otherwise difficult to transport and discard. In 1993, law professor David Voigt bluntly stated, "As the industry grew, the improper disposal of hazardous waste by maquiladoras turned much of the border between Mexico and the United States into an environmental disaster zone" (Voigt, 1993, 323).

Toxic materials used in a variety of industries along the border include arsenic, cadmium, lead, mercury, xylene, ethylbenzene, methylene chloride, acetone, toluene, and more. In light industries producing electronics, for example, large quantities of solvents and caustics were used. Concerns arose over unlawful dumping and the smuggling of hazardous wastes into Mexico for illegal disposal. Law scholar Elizabeth C. Rose stated in 1989, "When U.S. citizens and corporations consider nations like Mexico as their backyard there is a danger that they will also consider it as their outhouse" (Rose, 1989, 244).

Not only was hazardous waste a blight in the factories themselves but was particularly dangerous for those living in the Mexican towns where maquiladoras were located and in American colonias nearby. Praise for the maquiladoras for providing thousands of jobs along the border came at the expense of healthy settings for workers, poor living conditions, safety on the job, and decent wages.

Such exploitation was not new, but it changed little in this emerging industrial setting. Particularly aggravating was the unwillingness of those running the maquiladoras to divulge information about toxic exposures and other worker safety issues. This often was done in the name of maintaining tight security at the plants. New laws found their way onto the books, but they rarely led to regular inspections. One source noted that "Maquiladora workers voice constant fears about their safety on the job." Exposure to a variety of toxic substances in several industries—and without proper protection—can cause cancer, reproductive problems, skin diseases, respiratory issues, vision problems, gastrointestinal and nervous disorders, headaches, and fatigue. Workers also risk injuries to extremities, circulatory and muscular problems, and may face high stress. Some factories which produce intense noise levels often did not provide ear plugs, and some factories were not equipped with emergency doors (Kamel and Hoffman, 1999).

Away from the factories, communities surrounding the maquiladoras were susceptible to a variety of ailments and sickness. A 1991 *US News and World Report* survey found canals and rivers containing raw sewage caused widespread gastrointestinal illness, hepatitis, and other health matters. More studies followed. In an environmental justice case study in 1999, Elyse Bolsterstein quoted the American Medical Association as stating that the United States–Mexico border in the last 30 years was "a virtual cesspool and breeding ground for infectious disease." She also wrote that birth defects were a problem on both sides of the border, and that a 1994 study showed maquiladora women workers gave birth to lower weight babies than those working in other industries (Bolterstein, 1999).

Women have borne the brunt of unhealthy and unsafe conditions in the maquiladoras and received 10 to 30 percent lower wages than men. In 1999, two of three workers in maquiladoras were women (many younger than 20), while only 28 percent of the workforce in non-maquiladora industries were women. Demand increased for male workers, but primarily in heavy industries. In electronics, which is an important sector of the border industries, women workers dominate. Aside from risks on the job, major concerns arose about the effects of women labor on family life in those communities (Williams and Homedes, 2001, 323; Bolterstein, 1999).

More recent studies (between 2008 and 2015) confirmed earlier findings but also provided some nuance. A 2008 study of Juarez suggested that "while *maquiladoras* are more often located in formally developed areas inhabited by a more affluent populace, within these developed areas, it is the relatively less affluent who live closest to the *maquiladoras*" and thus are exposed to significant environmental risk. With respect to children, it is "the youngest and most biologically vulnerable that bear the largest burden from transnational industrialization in this Southern city." This conclusion might not seem unusual for those in the Global North, but in Mexico the more affluent in most cities tend to live in the dense core because of service availability, with the poor in the periphery where services are limited (Grineski and Collins, 2008, 266).

In a 2015, study of Tijuana, patterns of environmental injustice diverge from Juarez. In Tijuana "generally economically better-off residents...are exposed to greater densities of industrial hazards, which is the opposite of what is usually found in the Global North." In addition, the study showed that the risk was high for more affluent female-headed households along with female maquiladora workers who choose to live near their places of employment (Grineski et al., 2015).

It was not until 1994 with the passage of the North American Free Trade Agreement (NAFTA) that many hoped that current border problems could be alleviated or at least improved: These included the enforcement of environmental laws, better working conditions, and decrease in the concentration of maquiladoras in close proximity to the border. NAFTA was envisioned as the first major free-trade agreement to establish an integrated continent-wide free-trade market for the United States, Canada, and Mexico. But, according to environmental policy expert Richard N.L. Andrews, it involved "not simply an abstract commitment to free trade but the visible consequences of job losses and pollution associated with 'maquiladora' industries operating just across the Mexican border" (Andrews, 2020, 376). Attempting to fix this, President Bill Clinton negotiated two "side agreements" to NAFTA dealing with environmental and labor protections before ratification of the major agreement in the Senate took place (1993).

NAFTA, however, remained controversial. It was blamed for job losses and wage reductions in U.S. communities. It also disrupted Mexico's agricultural economy whose farmers now faced competition from imports such as subsidized American corn. And its raised serious question about the extent to which it helped resolve many environmental problems. Several years later President Donald Trump's decision to renegotiate NAFTA created a variety of additional uncertainties. His determination to build a wall between Mexico and the United States, among other things, raised several red flags about the environmental implications of such a structure (Andrews, 2020, 376–377; Carruthers, 2008, 138–141; James, 2019).

A story in the *Los Angeles Times* on June 30, 1997, asserted: "The NAFTA environmental program, drawn up to win over congressional opponents of free trade in this hemisphere, was never intended as an all-out war on border pollution. The agreement did not create tough new police power to make up for years of regulatory neglect and shoddy enforcement of environmental laws on both sides of the border. NAFTA did not mandate a crackdown on polluters or a cleanup of existing pollution" (Clifford and Sheridan, 1997). The story conceded, however, that NAFTA did help communities build better water and sanitation systems.

Yet, NAFTA gave priority to boosting economic growth through the maquiladoras and kept low barriers to trade. Thousands of people still rushed to the border to take the poor wages and ended up living in substandard housing in the process. NAFTA's stiffest critics contend that the agreement did the opposite of what the countries agreed to do with respect to environmental protection (especially along the United States–Mexico border), exacerbating existing problems through its economic policies (McDowra, 2020; Grineski et al., 2015).

Economic success for businesses operating along the border has not been matched by significant environmental improvement. Since the risks from water scarcity, water pollution, hazardous wastes, and air pollution have been experienced all along the border disproportionately by workers and the poor (most of whom are people of color) our story is one of vast environmental injustice. By 1990, 88 percent of the maquiladoras were located along the arid northern Mexico's border communities, with some beginning to move inland. In the 1980s the distribution had begun to change.

The industries had been historically located along the northeastern border in places like Juarez, Laredo, Reynosa, and Matamoros. As business boomed, the number of maquiladoras in the northwest increased in Tijuana, Tecate, Mexicali, Nogales, and Agua Prieta. Ciudad Juarez became the most important maquiladora center in Mexico. By 1987, it accounted for 20 percent of all maquildoras and 35 percent of the workforce. Tijuana ranked second, Matamoros third, Mexicali fourth, followed by Nogales, Nuevo Laredo, and Reynosa, all providing a wide variety of products (Sanchez, 1990, 164–165). By 1999, Tijuana had the most plants by far (735), but Juarez had the most employees (218,000). In the interior, Monterey was the leader with 132 plants and 57,000 employees (Williams and Homedes, 2001, 322).

The maquiladoras in all of their respective communities shared some forms of untenable pollution and risks, and many of those demonstrated acute environmental injustices. In Matamoros, in the northeastern state of Tamaulipas,

FINSA industrial park had one time been called by *Time* magazine "Love Canal in the making." Between 1988 and 1992, Matamoros became notorious when a cluster of birth defects and 30 cases of anencephaly (babies who were born with partially formed brains or skulls) were diagnosed in the Brownsville/Matamoros area. Prosperity from the maquiladoras came slowly there, with the poverty rate at 50 percent in Brownsville in 1985, and chronic housing shortages in Matamoros. The anencephaly incidents drew lawsuits in 1993 but were settled out of court two years later. In 1993–1994 more anencephaly cases showed up in Del Rio, Texas, and Tijuana (Kamel and Hoffman, 1999). Ultimately, several of the largest polluting industries in Matamoros left the area, but problems remained for those who stayed.

North of Matamoros in Ciudad Juarez, with the largest maquiladora labor force, air pollution has been the major problem, but not only attributable to the maquiladoras. Juarez has experienced several health and safety issues— chemical releases causing blindness and child deaths—in its factories. Not until the summer of 2000 did the first sewage treatment plant become operational, with another under construction (McDowra, 2020; Williams and Homedes, 2001, 328).

Mexicali, the capital of Baja California, became a dump for all kinds of waste, including old appliances, electronic waste, and used tires—some from the U.S. side of the border. The New River (Rio Nuevo) that flows from Mexicali to Calexico in the United States is considered the most polluted waterway in North America. Everything from pesticides and toxic metals to noxious bacteria have found their way into the river. As typical, information about pollution in Mexicali is primarily indirect. Factories in Mexicali— mostly America, but also Mexican, South Korean, Japanese, and European— vary from gigantic operations to smaller assembly plants.

Mexican government agencies have done little to monitor pollution or to provide effective oversight of the plants, especially concerning hazardous wastes. There has been little effort there to initiate a clear policy of regional development other than to attempt to boost employment in the maquiladoras. Mexicali has some of the worst air pollution in Mexico with high rates of asthma and respiratory deaths (James, 2019; McDowra, 2020; Sanchez, 1990, 163).

Tijuana, to the far west in Mexico, suffers many—if not all—of the environmental problems of the other border cities and towns. Clean-up has occurred at a snail's pace despite pressure from cross-border activist groups. Tijuana has had an ongoing regional water supply problem, augmented by dependence on water imports. Tijuana and Tecate began importing water in the mid-1990s, and in 2000 imports were 55.9 percent of water consumption in those cities (Stromberg, 2002, 35). In some cases, storage drums for toxic chemicals were used as drinking-water containers.

Fragile climate conditions in Tijuana have exacerbated erosion, flooding, and landslides over a substantial area because of the transborder interdependence of the drainage basins. Water pollution problems impact both sides of the border. Tijuana built its first sewage treatment plant in about 1990, but millions of gallons of untreated sewage and chemicals continued to run daily in the Tijuana River. The river was unfit for any use other than transporting the toxic stew.

Cross-border water pollution to this day has gotten worse rather than abated. The story is repeated elsewhere: rapid population growth, aging wastewater infrastructure, inadequate garbage collection, and factories that dump waste illegally. In February 2017 the deteriorating situation in Tijuana was graphically exposed when a winter storm cracked sewer pipes and man-hole covers spewing 200 million gallons of untreated sewage across the border and into the open sea. The toxic trail spread all the way to Coronado 16 miles northward up the coast (Tory, 2018).

A celebrated case of maquiladora pollution—which political scientist David V. Carruthers called "one of the most visible symbols of NAFTA's institutional failure"—involved the *Metales y Derivados* smelting facility at the edge of Tijuana's Otay Mesa industrial park (which opened in 1972). For more than 20 years, residents of Colonia Chilpancingo (in a canyon 150 yards below the industrial park) complained about the pollution. One study in 1990 found lead levels in the river below the colonia 3,000 times higher than U.S. standards, and cadmium 1,000 times higher. Finally, the plant was shut down in 1994 when the owners failed time and again to comply with waste disposal laws.

This was an unusual end to a border industry, yet the waste from the plant was never properly treated nor moved to a safer location. It simply leaked through containers, seeped into the ground, and contaminated the community water supply. Chilpancingo citizens formed the *Colectivo Chilpancingo Pro Justica Ambiental* in 2002, and finally in 2005, the Mexican government signed an agreement for a comprehensive cleanup. By the end of that year above-ground hazardous waste—at least—had been removed (Carruthers, 2008, 138–140; Bolterstein, 1999).

According to Carruthers:

> The US-Mexico border is an enigmatic place where the local and the global collide. It is at once prosperous and poor, urban and rural, Anglo American and Latin American. First World and Third World. In a few places do we see in such stark terms the unevenness with which the modern global economy parcels out costs and benefits. Border residents feel the environment and social contradictions of global development, North and South, with great intensity.
>
> (Carruthers, 2008, 137)

The maquiladoras have become a big part of that contradiction—wealth and economic power for corporations and governments, jobs for working-class Mexicans, but health and environmental risks falling disproportionately on those in the factories and in the border communities. Speaking about life along the border in an August 1990 issue of *Tulsa World,* an article noted, "Each day millions of gallons of raw sewage flow into open canals, called aguas negras, that runs intermittently along the 1,000-mile border between Texas and Mexico. Children play along the sides of the canals, where they contract the infectious diseases that are now rampant in the numerous South Texas slums, called colonias" (Maquiladora' Pollution U.S. Problem, 1990).

What was true for those Texas colonias was true for communities on both sides of the border. Such problems are still with us.

Notes and Further Reading

Andrews, Richard N.L. (2020) *Managing the Environment, Managing Ourselves: A History of American Environmental Policy* (New Haven, CT: Yale University Press, 3rd ed.).

Bolterstein, Elyse (1999) "Environmental Justice Case Study: Maquiladora Workers and Border Issues," online, http://websites.umich.edu/~snre492/Jones/maquiladora.htm.

Carruthers, David V. (2008) "Where Local Meets Global: Environmental Justice on the US-Mexico Border," in David V. Carruthers, ed., *Environmental Justice in Latin America: Problems, Promise, and Practice* (Cambridge, MIT Press), 137–160.

Clifford, Frank and Mary Beth Sheridan (1997) "Borderline Efforts on Pollution," *Los Angeles Times,* June 30.

Grineski, Sara E, et al. (2015) "Environmental Injustice Along the US-Mexico Border: Residential Proximity to Industrial Parks in Tijuana, Mexica," Environmental Research Letters 10, *IOPScience,* online, https://iopscience.iop.org/article/10.1088/1748–9326/10/9/095012.

Grineski, Sara E. and Timothy W. Collins (2008) "Exploring Patterns of Environmental Injustice in the Global South: *Maquiladoras* in Ciudad Juarez, Mexico," *Population and Environment* 29: 247–270.

James, Ian (2019) "This Mexican City Was Transformed by Factories. Its People Pay a Heavy Price," *Desert Sun,* December 15.

Kamel, Raechel and Anya Hoffman (1999) "Health and Environmental Issues: An Overview," *The Maquiladora Reader: Cross-Border Organizing Since NAFTA,* June 30, *Corpwatch,* online, https://www.corpwatch.org/article/health-and-environmental-issues.

Kotvis, Jill A. (1996) "NAFTA's Impact on the Environmental Management of the U.S./ Mexico Border Lands," *Law and Business Review of the Americas* 2: 50–101.

McDowra, Brandon (2020) "The Story of the Maquiladoras: Pollution, Corruption, and Destruction," December 9, storymaps.arcgis, online, https://storymaps.arcgis.com/stories/daed375a88ac430cba97a99e5a96d367.

Maquiladora' Pollution U.S. Problem (1990) *Tulsa World,* August 21.

Rose, Elizabeth C. (1989) "Transboundary Harm: Hazardous Waste Management Problems and Mexico's Maquiladoras," *International Lawyer* 23: 223–244.

Saldana, Lori (2007) "Tijuana's Toxic Waters," *NACLA* (North American Congress on Latin America), September 25, online, https://nacla.org/article/tijuana%27s-toxic-waters.

Sanchez, Roberto (1990) "Health and Environmental Risks of the Maquiladora in Mexicali," *Natural Resources Journal* 30 (Winter): 163–186.

Stewart, Alison L. (1992) "The Environmental Implications of North American Free Trade," *Environmental Law and Policy Journal* 15 (January): 3–11.

Stromberg, Per (2002) *The Mexican Maquila Industry and the Environment; An Overview of the Issues* (Estudios y Perspectivas, United Nations Publications).

Tory, Sarah (2018) "This Cross-Border Town Has faced Toxic Pollution for Years. Now Its residents Are Fighting Back." *Mother Jones*, December 9.

Voigt, David (1993) "The Maquiladora Problem in the Age of Nafta: Where Will We Find Solutions?" *Minnesota Journal of International Law* 2: 323–357.

Williams, Darryl M. and Nuria Homedes (2001) "The Impact of the Maquiladoras on Health and Health Policy along the U.S.-Mexico Border," *Journal of Public Health Policy* 22: 320–337.

25 To Frack or Not To Frack in Mexico

Fossil fuels and fresh water don't mix, just as oil and sea water don't mix. The practice of fracking in capturing oil and natural gas depends heavily on a sizeable supply of water in the process. Ultimately, the used water mixed with chemicals becomes unsuitable for human consumption and other uses. In the case of offshore oil drilling the medium of water—albeit salt water—is threatened in a different manner. Nevertheless, putting oil and water together has serious environmental consequences.

The introduction of large-scale fracking in the United States propelled the country into the front lines of oil and gas producers in the twenty-first century just as traditional extraction techniques raised America to its leadership role in the petroleum industry many decades earlier. With its vast underground and offshore resources, Mexico seemed likely to join the United States as a significant net exporter of energy in the near future, with fracking at the heart of the new exploration and retrieval of oil and gas. That promise, however, got mired in political, social, environmental, and economic battles that are unresolved to this day.

Hydraulic fracturing (fracking, fracing, or fraccing) is "a stimulation process" used to extract natural gas (and oil) from deep reserves below the ground (5,000–8,000 feet) which were previously inaccessible by traditional techniques. These reserves are found in shale and other forms of "tight" rock (impermeable rock formations). Fracking involves pumping water, chemicals, and proppant (often sand slurry and other materials) into a well at high pressure, and in the process fracturing the surrounding rock formation to open a passage through which the natural gas (oil) can flow. The chemicals utilized, among other things, are benzene, gelling agents, crosslinkers, friction reducers, biocides, and even diesel fuel. The carrying fluid (water and chemicals) can flow to the surface along with the gas. In most cases, only about 20 to 40 percent of the carrying fluid flows to the surface, while the rest remains in the ground. Early fracking technology used between 20,000 and 80,000 gallons of water per well, but advanced techniques use several million gallons of water plus 75,000–320,000 pounds of proppant per well (EEC Environmental, 2021).

Since the earliest discoveries of oil in the United States, using dynamite or nitroglycerin in a well hole (sometimes illegally) helped increase production.

DOI: 10.4324/9781003041627-34

Modern fracking dates back to 1862 when Lt. Colonel Edward A.L. Roberts observed how artillery affected narrow water channels. He applied these observations to developing an "exploding torpedo" (patented in 1865) that could be lowered into an oil well and then detonated. With the rock shattered, he could pump in water thus increasing oil flow. This became known as "shooting the well," and was used in Pennsylvania, New York, Kentucky, West Virginia, and elsewhere. In the 1930s, injecting a nonexplosive fluid of acid was tried to stimulate a well.

In the 1940s, high-pressure blasts of liquids largely replaced explosives—thus hydraulic fracturing. The first commercial application of fracking took place near Duncan, Oklahoma. Key changes in the twenty-first century include adding slickwater (a mix of water, sand, and chemicals—more than 1,000 chemicals were used) to make the fluid less viscous and using fracking alongside horizontal drilling (and later 3D seismic imaging) to reach more of the rock formation. With greater financing, approximately 1 million wells in the United States alone were fractured between 1940 and 2014 (Denchak, 2019; Wells and Wells, 2007; Montgomery and Smith, 2010, 26–28; Morton, 2013).

In the 1960s, the price of natural gas and the demand for the product were low. As a result, very few "tight gas plays" as they were called were economical enough to develop. In the 1970s, as prices increased and fracking technology improved, more numerous wells were developed in South and East Texas, the Mid-Continent, and the Rocky Mountains in the United States. In the 1980s and 1990s, research funding from public and private sources led to better knowledge about formations and the fracturing process itself. The technique, however, was not used on a large scale until 2003 when companies expanded natural gas exploration (emphasizing shale formations) in Texas, Pennsylvania, West Virginia, Wyoming, Utah, and Maryland (Holditch, 2007, 116; EEC Environmental, 2021).

Since 2014, fracking wells were the majority of new oil and natural gas wells in the United States. In 2016, almost 70 percent of the country's 977,000 producing wells were subjected to the fracking process and horizontal drilling. Freelance writer and editor Melissa Denchak stated, "The fracking boom is largely credited with making the United States the top producer of natural gas and crude oil in the world—a trend expected to continue as fracking becomes more efficient...and enables access to more of the country's fossil fuel reserves." Texas is the top producer of crude oil and natural gas; North Dakota is second for crude oil, and Pennsylvania is second for natural gas. Some states and cities have resisted developing their reserves, however. In 2015, New York became the first state with substantial natural gas reserves to prohibit fracking for public health and environmental reasons (Denchak, 2019).

The success of hydraulic fracturing in the U.S. cannot be understated: changing the business landscape and helping to sustain the prominence of fossil fuels in the energy market. But such success with fracking came with a price as growing evidence exposed serious threats to human health and to the environment in general.

The insatiable demand for water in the fracking process is particularly noteworthy. The fluid mixture utilized is comprised of as much as 97 percent water. The U.S. Geological Survey estimated that fracking in recent years consumes between 1.5 and 16 million gallons of water per well, depending on the type of well drilled and the rock formation where fracking took place. Normally, fresh water is taken from surface and groundwater in the area.

The industry has stated that it has increased efforts to use nonpotable water when possible, and that compared to other industries fracking uses small amounts of water. This is playing numbers games. Water for fracking may be "negligible" measured against the aggregate amount of water utilized across the nation for various purposes, but depletion of fresh water has to be taken into account on a local basis. If, for example, a well(s) is drilled in proximity to a small town dependent on a nearby water source, the impact can be quite severe. In the arid Permian Basin in West Texas water used for fracking between 2011 and 2016 increased by about 770 percent (Denchak, 2019).

Aside from the volume of water used, fracking may cause contamination of aquifers. Fluids for fracking leave a very large amount of wastewater which, in most cases, remains untreated. Spills and leaks can occur throughout the process from the transportation and storage of chemical additives to mixing and pumping the fluids. In many cases, the potential health risks of the chemicals for humans remain unknown but ominous.

A U.S. Environmental Protection Agency (EPA) study in 2004 found that fracking did not pose a threat to underground drinking water supplies—a finding that the industry embraced. In the following year, the George Bush administration's Energy Policy Act (2005) exempted fracking from the Safe Drinking Water Act. In 2010, however, the Awareness of Chemicals Act, a bill to amend the Safe Drinking Water Act, called for the repeal of the 2005 exemption for fracking. The EPA then set out to further study the relationship between fracking and drinking water and reported in 2015 that fracking may impact drinking water in some cases, thus changing its 2004 stance. In 2021, it confirmed its 2015 findings.

Nearby communities face health threats also from air pollution produced during the fracking process. This includes flaring for testing or safety purposes, venting gases directly into the air, or through unintentional leaking of contaminants. Natural gas is composed primarily of methane, a very potent greenhouse gas that if flared, vented, or leaked can directly impact climate change. In addition, the building of elaborate infrastructure for drilling operations can harm forest and rural landscapes, damage animal habitats, and elevate noise levels. The fracking process is known to produce micro-earthquakes (as a part of fracking itself), and in several cases larger earthquakes in specific regions have been attributed to fracking operations. In 2021, several stories appeared in a variety of environmental publications claiming that

Figure 25.1 Fracking operation. Joshua Doubek. Creative Commons Attribution-Share Alike 3.0. Wikipedia.

fracking also has led to millions of gallons of toxic waste being dumped into the Gulf of Mexico.

The response of the industry is to deny many charges of adverse environmental impacts, claim steady improvement in available technology, and tout the economic value of fracking. In 2010, oil and gas attorney Wes Deweese argued that "Hydraulic fracturing is a safe, environmentally sound oil and gas recovery method. It is also essential to meeting America's growing demand for natural gas." He added that fracking is "effectively regulated by states," but proposed national regulations would only produce a "one-size-fits-all approach," "unnecessarily transferring to the federal government the regulation of an industry practice that has been effectively regulated by states" (Deweese, 2010, 1–2).

As fracking became more widespread in the 2010s, controversy grew rapidly in the United States and internationally. Opponents continued to present evidence about the environmental risks of fracking and claimed, unlike industry spokespersons, that fracking remained acutely underregulated. In several cases, state-level oil and gas agencies do not require companies to report the volumes or names of chemicals injected during fracking, nor do they conduct sampling to determine what happened to the fracturing chemicals after use.

In 2021, the Interstate Delaware River Basin Commission banned fracking in the watershed of New York, New Jersey, Pennsylvania, and Delaware (adjacent to the Marcellus shale basin). Vermont, Maryland, and Washington also banned fracking, but those states have few, if any, reserves. California is prepared to ban fracking by 2024. Despite persistent concerns, fracking is still big business in the United States, although oil and gas prices fell sharply during the early stages of the coronavirus pandemic (and later rose again).

Commercial oil production in Mexico began in Veracruz in 1868. In 1938, President Lazaro Cardenas expropriated foreign oil companies leading to the formation of *Petroleos Mexicanos* (PEMEX) which was meant to conduct all petroleum exploration, exploitation, refining, and marketing in Mexico. From 1938 to December 20, 2013, Mexico regulated all petroleum production based on Article 27 of the Mexican Constitution.

Mexico, like the United States, offered great potential for natural gas and shale oil production. According to the Energy Information Administration in 2013, Mexico had estimated recoverable shale resources of 545 trillion cubic feet of natural gas and 13.1 billion cubic feet of oil and condensate. This gave Mexico the sixth-largest reserves of recoverable shale gas in the world, just behind China, Argentina, Algeria, the United States, and Canada (in that order). Mexico also was the eighth largest in the world with recoverable shale oil.

PEMEX had dabbled with the fracking method since the 1980s. More systematic exploration for potential drilling sites for shale oil and gas began in Mexico late in 2010 (some say earlier). Promising locations included the Chihuahua region (northwest), Sabinas-Burro-Picachos region (northeast), Tampico-Misantla (southeast), Veracruz (southeast), and the Burgos Basin. Located in northeastern Coahuila state, the Burgos Basin (part of the Texas Eagle Ford Shale area) accounts for as much as two-thirds of the recoverable shale gas in Mexico.

PEMEX made public its first test well for shale gas in Burgos Basin in April 2013. "The Burgos Basin shale gas fields, which spans the arid shrublands of Northern Tamaulipas Mexico, is a potential bonanza as big or bigger than the record-setting Eagle Ford shale play across the Rio Grande in South Texas," reported the North American Congress on Latin America. But after this effort, PEMEX drilled only a limited number of test wells. Despite the early success, it may have stopped the activity due to high costs and significant budget cuts in 2015 and 2016. A call for an estimated 40,000 new wells to take advantage of the shale resources was stymied by politics, security concerns, and importantly the lack of water (Vinson & Elkins, 2018).

A landmark federal law in 2013 appeared to change everything for Mexico's fracking future, which promised to fulfill the potential of one of the country's most promising resources. In December 2013, Mexico's Congress approved a constitutional amendment brought forward by President Enrique Pena Nieto that would reopen the country's oil industry to private and foreign investment. With the reforms, the state maintained ownership of subsoil hydrocarbon resources, but private companies could take ownership of resources once they were extracted. PEMEX would contract with private companies for shale and gas exploration. And the government would strengthen the regulatory entities that would oversee the new activity. Such changes were very difficult for PEMEX to accept, since it had been a state monopoly since the expropriation in 1938 (Hyatt, 2017, 696–697).

The law was meant to lure companies drilling for oil in the Gulf of Mexico and exploiting the Texas shale resources to Mexico. Government officials hoped the companies would bring new technology, technical and administrative expertise, and capital that was in short supply through PEMEX. The law also brought to an end total state control of the oil and gas industry in Mexico. "This is critical to the re-industrialization of North America," stated Javier Trevino, the head of the energy commission in the Congress' lower house. "Mexico needs to develop these resources, or else we'll be left behind" (*Washington Post*, April 19, 2014).

Pena desired raising Mexico's oil production to 3 million barrels a day by 2018, which was a 25 percent increase from 2013. Goldman Sachs estimated that development of the shale resources would require a massive $1.2 trillion in capital spending. Mexico appeared to have its greatest chance for success in the Burgos Basin where there already were 11,000 well permits on the U.S. side. There was some reticence, however, among companies in the Mexican states bordering the shale area in Texas and the Gulf of Mexico because of drug cartels operating there (such as the Zetas and Gulf Cartel). In 2013, PEMEX found 539 siphons along the pipelines in Tamaulipas. Between 2006 and 2012, incidents of gangs illegally tapping pipelines along the Texas-Mexico border increased more than six times (Corchado and Osborne, 2014; Mexican Drug Cartels, Quartz, 2014). In some cases, PEMEX geologists and survey crews needed military escorts while looking for new well sites.

In December 2014, the National Hydrocarbons Commission in Mexico announced its bidding guidelines for development blocks to private companies. Round One included five tenders—two for offshore, one for onshore fields, and another for unconventional deposits of shale (using fracking). There were stops and starts, and the response was modest since lower oil prices weakened bids, and Mexico had to improve terms it offered to bidders to get the results they hoped for. In March 2018, Mexico announced that it would auction development rights to shale in Burgos Basin. This would be the first time private oil companies had a chance to develop such sources in Mexico. As of June 2018, Mexico conducted nine oil and gas auctions, awarding about 90 contracts on offshore and onshore sites, but development crawled at a snail's pace (Vinson & Elkins, 2018).

Mexico had a rocky start in developing oil and gas sites under the new rules and faced an even more unsettling event in December 2018. Mexico's president-elect Andres Manuel Lopez Obrador announced that he would end the use of hydraulic fracturing once he entered office. The new populist leader asserted that because of environmental concerns, "We will no longer use that method to extract petroleum"—at least during his six-year term in office. Lopez Obrador, however, also wanted to increase conventional natural gas production in Mexico to offset the extensive importing of natural gas from the United States. Thomas Tunstall, Research Director at the University of Texas at San Antonio's Institute of Economic Development, believed that the proposed fracking ban was mostly symbolic and would have little immediate economic impact. He surmised that Mexico had substantial conventional reserves. "In addition," he added, "most of the unconventional/shale opportunities lie in North Mexico, which lacks significant infrastructure ... Best estimates are that any unconventional oil and gas production activity in Mexico is at least 5–10 years away, no matter what government policy is" (EcoWatch, 2018).

The president's apparent mixed signals on natural gas development made his country's energy policy fuzzy at best, since regulations allowing fracking in Mexico were not struck from the books. In office, Lopez Obrador also expressed sharp criticism of the 2013 energy law but stopped short of restoring the PEMEX monopoly. Some supported the president's pronouncement about a fracking ban, such as Environmental Minister Victor Toledo, who wanted to see legislation written to prohibit fracking.

Private business leaders such as Tania Ortiz, CEO of Sempra Energy's Mexico unit, held a commonly held view among proponents of fracking that Mexico's abundant shale reserves should not go undeveloped. Others believed fracking was inevitable, especially since the Federal government had no clear stance on banning fracking despite Lopez Obrador's public statements. Businessmen and political leaders also were watching what happened in Veracruz, where exploratory natural gas wells had been authorized for PEMEX, to see if the process was viable after all. But as of 2021, fracking in Mexico remains highly controversial, and a clear plan for unconventional oil and gas production employing it has yet to be settled.

In most respects, Lopez Obrador's stance on fracking reflected public anti-fracking sentiment among his constituents entangled with questions of local versus federal authority, environmental concerns, and access to water. With the changing energy policy of the Mexican government in 2013–2014, anti-fracking forces (such as the Chihuahua Citizen Alliance against Fracking) were mobilizing and calling for a moratorium or a complete ban of hydraulic fracturing. They also accused the government of suppressing national debate on the process, a debate that had been going on in the United States, other countries in Latin America, and throughout the rest of the world. Documents prepared by the Center for Human Rights and Environment in 2015 outlined a human rights assessment of fracking, which was reflected in the rhetoric of anti-fracking groups in Mexico.

In July 2014, the Mexican Alliance Against Fracking presented a petition with 14,000 names to the Mexican Congress and staged a protest in Mexico City attended by an estimated 60,000 people. On the local level activists were speaking out to educate communities about the possible negative impacts of fracking. After Lopez Obrador's election, protests groups such as the Environment Ministry supported a fracking ban and encouraged the development of more environmentally responsible ways of developing hydrocarbon extraction. Some groups took more direct action. In the northern state of Sonora, the Yaqui people blocked the construction of the private *El Oro-Guaymas* gas pipeline that left some wounded or killed.

An obvious reason for resistance to fracking was protecting the local community from environmental damage and health concerns. The same criticisms raised north of the border were echoed in Mexico. While the petroleum industry would deny such blanket charges, the Center for Human Rights and Environment declared:

> Despite the very recent arrival of intense fracking activity, the extraction procedure utilized by hydraulic fracturing has already been around long enough to leave a considerable environmental footprint. The most notorious documented fracking impacts are due to burst underground well pipe casings contaminating water aquifers, fugitive methane leaks from pressurized gas not only causing climate change, but creating nauseous clouds that make local communities sick, and the seepage of industrial effluents affecting surface water resources and other sensitive ecosystem resources.
>
> (Center for Human Rights and Environment, 2015, 6)

Concern in Mexico, as elsewhere, also focused on the lack of specific environmental regulations related to fracking and—especially in the case of Mexico—lax environmental regulations overall. In Mexico's north in particular concern over fracking-induced earthquakes increased after 2013. *The World* reported in 2014, "The vibrations grew more frequent into this spring, cracking the walls of some of the cinder block farmhouses. Between January and mid-April, there were a record 48 tremors across the state of Nuevo Leon…compared with two in the same period last year…" Ruperto De La Garza, an environmental impact consultant stated, "The one thing that has changed is the introduction of fracking. There is no other explanation" (Grillo, 2014).

Even more than in the United States, the twin issues of water and drought had a major impact on the future of fracking in Mexico. Attorney Jacqueline F. Hyatt stated, "Water catalyzed political change throughout Mexican history" from the water management of Tenochtitlan to the 1910 Revolution and beyond (Hyatt, 2017, 696). Looking toward the possibility of large-scale fracking in Mexico, Eileen Wray-McCann in *Water News* stated flatly, "Mexico needs investment and infrastructure to support hydraulic fracturing, but a great challenge, and an even greater concern, is access to water" (Wray-McCann, 2019).

Especially, but not exclusively, in north and central Mexico access to water had been and continued to be a chronic problem. In the location of the Eagle Ford shale formation (Burgos Basin), concern and conflict over water scarcity were well-known. In 2012, approximately 350,000 head of cattle had starved to death in Chihuahua because of the shortage of pastureland due to the lack of water (Estevez, 2014).

Water shortages created conflict, and fear of fracking made things worse. In 2015, concerns over water shortages brought on by drilling test wells led farmers in the Coahuila municipality of Jimenez to file a complaint with the Latin American Water Tribunal (an independent entity based in Costa Rica) about water shortages in their community which they attributed to fracking. The federal government had assured the locals that Jimenez was an area where fracking would not be allowed. But private energy companies and municipal politicians attempted to convince residents of the *ejidos* (an area of communal land use for farming) that fracking had financial benefits. Locals were resisting private company efforts to buy up their property, but the pressure was great (Navarro, 2016, 2).

Conagua (La Comision Nacional del Agua), Mexico's primary water agency, made the point that fracking would demand only a small percentage of the water allocated to industry (about five percent). The assumption being that this amount did not threaten other uses—an argument like the one used in the United States. As in the United States, fresh water used for fracking impacts specific communities, and any use does not take into account the polluting of freshwater sources utilized in the process or the potential risk to groundwater. In recent times, energy companies that were starting new projects bought water from those with existing water claims, such as farmers, but such efforts did not take into account what drought might do to those supplies or to the competing demand for them.

There is little doubt that water will prove to be one of the biggest—if not the biggest— tests to expanding fracking in Mexico. At the very least, the issue of water complicates conducting fracking business in Mexico where ownership to subsurface mineral rights rests with the government rather than the landowner; where the actions of local, state, and federal regulatory agencies are not coordinated; where state politics waivers between anti-fracking and the desire to secure new economic opportunities; and where security of resources is threatened by drug cartels.

In the last few years, immediate problems due to stagnating GDP growth, the global pandemic, low oil prices, possible recession, and perennial protests that all converged on the fracking debate in Mexico were aggravated by discussion of the possible long-term implications of climate change. Marcella Duenas, market engagement manager for S&P Global Platts, sums it up best, "To frack or not to frack, is the dilemma that lingers" (Duenas, 2019).

Notes and Further Reading

"A Brief History of Hydraulic Fracturing" (2021) *EEC Environmental*, online, https://eecenvironmental.com/a-brief-history-of-hydraulic-fracturing/.

Center for Human Rights and Environment (2015) *Human Rights and the Business of Fracking: Applying the UN Guiding Principles on Business and Human Rights to Hydraulic Fracturing*, Consultation Draft, August 20, 1–10.

Corchado, Alfredo and James Osborne (2014) "Eyes are on Mexico's Untapped Potential," *Dallas Morning News*, online, http://res.dallasnews.com/interactives/border_energy/.

Denchak, Melissa (2019) "Fracking 101," *Natural Resources Defense Council (NRDC)*, April 19, online, https://www.nrdc.org/stories/fracking-101.

Deweese, Wes (2010) "Fracturing Misconceptions: A History of Effective State Regulation, groundwater Protection, and the Ill-Conceived Frac Act," *Oklahoma Journal of Law and Technology* 6 (January): 1–32.

Duenas, Marcela (2019) "Mexico's Gas Dependence on US Pushes Politicians to Consider Fracking," *S&P Global Platts*, April 3, online, https://www.spglobal.com/platts/en/market-insights/blogs/natural-gas/040319-mexicos-gas-dependence-on-us-pushes-politicians-to-consider-fracking.

Estevez, Dolia (2014) "Fracking: Could Mexico's Water Scarcity Render Its Energy Sector Reforms Self-Defeating?" Forbes (June 11), online, https://www.forbes.com/sites/doliaestevez/2014/06/11/fracking-could-mexicos-water-scarcity-render-its-energy-sector-reforms-self-defeating-2/?sh=3f37bff754ea.

Gold, Russell (2014) *The Boom: How Fracking Ignited the American Energy Revolution and Changed the World* (New York: Simon & Schuster).

Grillo, Ioan (2014) "Fracking Near the Texas Border Has Northern Mexico Trembling," *The World*, April 29, online, https://www.pri.org/stories/2014-04-29/fracking-near-texas-border-has-northern-mexico-trembling.

Holditch, Stephen A. (2007) "Hydraulic Fracturing: Overview, Trends, Issues," *Drilling Contractor* July/August: 116–117.

Hyatt, Jacqueline F. (2017) "The Energy-Water Nexus: Water Regulation in the Wake of Mexico's Hydrocarbon Reform," *Houston Journal of International Law* 39 (July 19): 695–722.

Mazur, Allan (2014) "How Did the Fracking Controversy Emerge in the Period 2010–2012?" *Public Understanding of Science* 23: 1–16.

"Mexico's Drug Cartels Are Standing in the Way of the Fracking Bonanza" (2014), *Mexico's Drug Cartels*, February 20, online, https://qz.com/178986/mexicos-drug-cartels-are-standing-in-the-way-of-a-fracking-bonanza/.

Mitchell, Timothy (2011) *Carbon Democracy: Political Power in the Age of Oil* (London: Verso).

Montgomery, Carl T. and Michael B. Smith (2010) "Hydraulic Fracturing: History of an Enduring Technology," *Journal of Petroleum Technology* 62 (December): 26–32.

Morton, Michael Quentin (2013) "Unlocking the Earth: A Short History of Hydraulic Fracking," *GEOExPro* 10, online, https://www.geoexpro.com/articles/2014/02/unlocking-the-earth-a-short-history-of-hydraulic-fracturing.

Navarro, Carlos (2016) "Mexican Government Set to Award Fracking Contracts in Northeast," University of New Mexico, Latin American & Iberian Institute, October 5.

"Newly Elected President of Mexico to Ban Fracking" (2018) *EcoWatch*, August 02, online, https://www.ecowatch.com/mexico-bans-fracking-2592125998.html.

Vinson & Elkins (2018) "Mexico: Global Fracking Resources," Shale & Fracking Tracker, September, online, https://www.velaw.com/shale-fracking-tracker/resources/mexico/.

Wells, B.A., and K.L. Wells (2007) "Shooters-A 'Fracking' History," *American Oil & Gas Historical Society*, September 1, online, https://aoghs.org/technology/hydraulic-fracturing/.

Wray-McCann, Eileen (2019) "What's Up with Water—Mexico, Aproved Exploratory Fracking Wells in Pursuit of Energy Independence," *Water News* March 4, online, https://www.circleofblue.org/2019/world/whats-up-with-water-mexico-approved-exploratory-fracking-wells-in-pursuit-of-energy-independence/.

Yergin, Daniel (2011) *The Quest: Energy, Security, and the remaking of the Modern World* (New York: Penguin Press).

Postscript: Climate and Water

This book has dealt with water—in many eras, places, and circumstances—as an essential element in North American environmental history. Some of the episodes have briefly intimated how climate change is beginning to influence water use, water scarcity, and water flow. Into the future, climate change is certain to have a much bigger impact on water events of all kinds.

Environmental historian Anthony Penna has observed that "Scientists identify three basic causes of climate change: the exchange of energy by the oceans and atmosphere, fossil fuel emissions, and solar energy." Penna added, "Although the specific impacts of a warming global climate system on various populations in different parts of the world remain a subject of ongoing research, climatologists point out that during the current warming phase we can expect the physical properties of CO_2 to contribute to more rainfall, higher atmospheric and oceanic temperatures, more clouds and higher wind velocity. The biological effects of CO_2 may also contribute to longer growing seasons in temperate climates…With an unequal distribution of rainfall, however, formerly well-watered environments may become stricken by long uninterrupted periods of drought." By using the planet's atmosphere as a "carbon sink," we also have experienced more Arctic sea ice melts, sea-level rise, changes in aquatic diversity, flooding, massive wildfires, erosion of coastlines, threats to our water-supply systems, and stronger hurricanes (Penna, 2010, 287–289: McNeill and Engelke, 2014, 68–71, 86, 91–97). A 2008 Intergovernmental Panel on Climate Change technical paper on climate change and water stated, "Climate, freshwater, biophysical and socio-economic systems are interconnected in complex ways. Hence, a change in any one of these can induce a change in any other" (Melosi, 2011, 201).

While we most often think about climate change in terms of carbon emissions into the atmosphere, many experts and commentators recognize that water bears a great burden in an environment of rising temperatures. A United Nation's report stated that "Water is the primary medium through which we feel the effects of climate change." For low-income people, in particular, water availability is and will continue to become "less predictable" in many places, and the incidence of drought, increased flooding, and disruption of sanitary services will impair people's health (United Nations, 2021). Thus, as the International Union for Conservation of Nature (IUCN) explained, "Water and weather, the delicate balance between evaporation and precipitation, is the primary cycle through which climate change is felt" (International Union for Conservation of Nature, 2015, 1).

The U.S. Environmental Protection Agency (EPA) came to the conclusion that climate change "is changing our assumptions about water resources." It emphasizes that warming temperatures are altering the hydrologic cycle, changing the "amount, timing, form, and intensity" of precipitation, affecting the flow of water in watersheds, and modifying the quality of aquatic and marine environments. Such outcomes are influencing EPA's programs to protect water quality, public health, and safety (EPA, 2021).

Figure 26.1 Protester at a rally dealing with climate change. Ivan Radic. Creative Commons Attribution 2.0. Wikimedia.

Recently, California's Department of Water Resources stated, "Climate change is already impacting water and other resources in California, and will continue to do so as California's population and demand for water increases." Current indicators include reduced snowpack in the mountains, weather extremes (such as wildfires), and sea-level rise. The concern is increasing over the long-term impacts on the state's agricultural economy (California, 2021). Such observations about climate change, however, have not been restricted to California. Environmentalist Bill McKibben, in writing about the Great Lakes—and especially Lake Erie—maybe too glibly remarked, "Even areas that we're used to thinking of as ruined may be ruined in new and interesting ways…They have recovered somewhat, but a change in climate may subject them to unprecedented stresses (which is like subjecting the South Bronx to unprecedented decay)" (McKibben, 2006, 102–103).

Declining water quality is another very real consequence of climate change. Water temperature in streams, lakes, rivers, and reservoirs increases as air temperature rises. This can lead to lower levels of dissolved oxygen in the water with more stress on fish and other aquatic animals and plants. More intense rainfall means greater amounts of runoff (containing sediments, disease pathogens, pesticides, and herbicides) into watercourses resulting in greater pollution. Groundwater is also susceptible to larger amounts of polluted material entering aquifers.

Given its numerous impacts, it is no wonder that some people are calling the climate crisis a water crisis. Grim as the predictions might be, such projections about future calamities are beginning to be taken more seriously. As to whether we will respond effectively to these threats remains to be seen. With respect to our exploration of water in environmental history, however, the climate crisis demonstrates once again how indispensable water is to life and how often it is a medium for understanding living itself.

Notes and Further Reading

California Department of Water Resources (2021) "Climate Change and Water," online, https://water.ca.gov/Programs/All-Programs/Climate-Change-Program/Climate-Change-and-Water (accessed November 10).

International Union for Conservation of Nature (2015) "Water and Climate Change: Building Climate Change Resilience through Water Management and Ecosystems," November, 1–2, online, https://www.iucn.org/sites/dev/files/import/downloads/water_and_climate_change_issues_brief.pdf.

Journal of Water & Climate Change (2010-).

Kress, John W. and Jeffrey Stine, eds. (2017) *Living in the Anthropocene: Earth in the Age of Humans* (Washington, D.C.: Smithsonian Books).

McKibben, Bill (2006) *The End of Nature* (New York: Random House, new ed.).

McNeill, John and Peter Engelke (2014) *The Great Acceleration: An Environmental History of the Anthropocene since 1945* (Cambridge, MA: Belknap Press).

Melosi, Martin V., ed. (2011) *Precious Commodity: Providing Water for America's Cities* (Pittsburgh, PA: University of Pittsburgh Press).

Penna, Anthony N. (2010) *The Human Footprint: A Global Environmental History* (Chichester, UK: Wiley-Blackwell).

United Nations (2021) "Water and Climate Change," online, https://www.unwater.org/water-facts/climate-change/ (accessed November 10).

U.S. Environmental Protection Agency (2021) "Addressing Climate Change in the Water Sector," online, https://www.epa.gov/climate-change-water-sector (accessed November 10).

Index

Note: Italicized page numbers refer to figures.